MEI structured mathematics

Pure
Mathematics 2

VAL HANRAHAN
ROGER PORKESS
PETER SECKER

Series Editor: Roger Porkess

MEI Structured Mathematics is supported by industry:
BNFL, Casio, GEC, Intercity, JCB, Lucas, The National Grid Company,
Sharp, Texas Instruments, Thorn EMI

Hodder & Stoughton

A MEMBER OF THE HODDER HEADLINE GROUP

Acknowledgements

We are grateful to the following companies, institutions and individuals who have given permission to reproduce photographs in this book. Every effort has been made to trace and acknowledge ownership of copyright. The publishers will be glad to make suitable arrangements with any copyright holder it has not been possible to contact.

The Image bank © P.M. Miller (51)

Roddy Paine Photographer (93)

The University of Cambridge Local Examinations Syndicate, Oxford and Cambridge Examinations Council and the Schools Maths Project accept no responsibility for the answers to questions taken from past question papers contained in this publication.

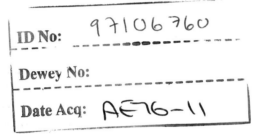
Orders: please contact Bookpoint Ltd, 39 Milton Park, Abingdon, Oxon OX14 4TD. Telephone: (44) 01235 400414, Fax: (44) 01235 400454. Lines are open from 9.00 - 6.00, Monday to Saturday, with a 24 hour message answering service. Email address: orders@bookpoint.co.uk

British Library Cataloguing in Publication Data
A catalogue record for this title is available from The British Library

ISBN 0 340 57302 3

First published 1995
Impression number   14   13   12   11   10   9   8   7   6   5   4
Year   2004   2003   2002   2001   2000   1999   1998

Typeset by Multiplex Techniques Ltd.
Printed in Great Britain for Hodder & Stoughton Educational, a division of Hodder Headline Plc, 338 Euston Road, London NW1 3BH by Scotprint Ltd, Musselburgh, Scotland.

MEI Structured Mathematics

Mathematics is not only a beautiful and exciting subject in its own right but also one that underpins many other branches of learning. It is consequently fundamental to the success of a modern economy.

MEI Structured Mathematics is designed to increase substantially the number of people taking the subject post-GCSE, by making it accessible, interesting and relevant to a wide range of students.

It is a credit accumulation scheme based on 45 hour components which may be taken individually or aggregated to give:

3 components   AS Mathematics
6 components   A Level Mathematics
9 components   A Level Mathematics + AS Further Mathematics
12 components   A Level Mathematics + A Level Further Mathematics

Components may alternatively be combined to give other A or AS certifications (in Statistics, for example) or they may be used to obtain credit towards other types of qualification.

The course is examined by the Oxford and Cambridge Schools Examination Board, with examinations held in January and June each year.

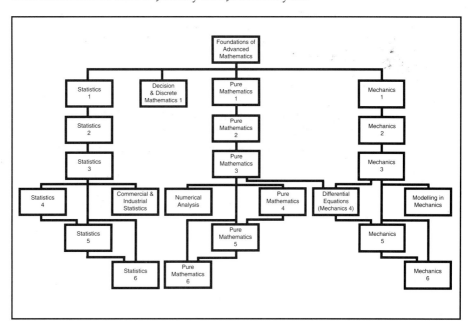

This is one of the series of books written to support the course. Its position within the whole scheme can be seen in the diagram above.

Mathematics in Education and Industry is a curriculum development body which aims to promote the links between Education and Industry in Mathematics, and to produce relevant examination and teaching syllabuses and support material. Since its foundation in the 1960s, MEI has provided syllabuses for GCSE (or O Level), Additional Mathematics and A Level.

For more information about MEI Structured Mathematics or other syllabuses and materials, write to MEI Office, 11 Market Street, Bradford-on-Avon BA15 1LL.

Introduction

This is the second in a series of books written to support the Pure Mathematics Components in MEI Structured Mathematics. These books are also suitable for an independent course in the subject. *Pure Mathematics 1 and 2* cover the pure subject core for AS Mathematics, the first three books that for A Level. Throughout the series the emphasis is on understanding rather than on mere routine calculations, but the various excercises do provide plenty of scope for practising basic techniques.

Some parts of this book lay the foundation for future work, with the introduction of major new ideas: functions, sequences and series, logarithms and exponentials. Other parts build on work in *Pure Mathematics 1*, particularly the calculus chapters which extend the range of functions which you are able to differentiate and integrate, and introduce you to the use of substitution. The final chapter covers several numerical techniques for the solution of equations which you should find very useful.

We would like to thank Alan Sherlock for the ideas which he contributed to this book, and all those who have helped us by reading the text, in many cases trialling it with their students. Finally we are grateful to the various examination boards who have given permission for their past questions to be included in the exercises.

Val Hanrahan, Roger Porkess and Peter Secker.

Contents

Indices

"A good notation has a subtlety and suggestiveness which at times makes it seem almost like a live teacher."

Bertrand Russell

The graph below refers to the moons of the planet Jupiter. For six of its moons, the time, T, that each one takes to orbit Jupiter is plotted against the average radius, r, of its orbit. (The remaining ten moons of Jupiter would be either far off the scale or bunched together near the origin).

The curves $T = kr^1$ and $T = kr^2$ (using suitable units and constant k) are also drawn on the graph. You will see that the curve defined by Jupiter's moons lies somewhere between the two.

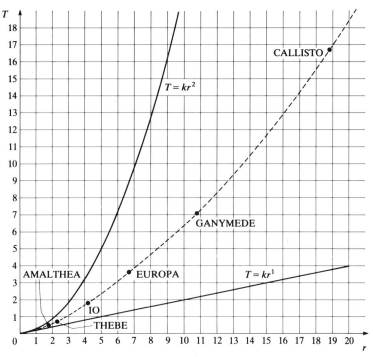

How could you express the equation of the curve defined by Jupiter's moons?

Negative and fractional indices

You may have suggested that the equation of the curve defined by Jupiter's moons could be written as $T = kr^{1\frac{1}{2}}$. If so, well done! This is correct, but what would a power, or index, of $1\frac{1}{2}$ mean?

Before answering this question it will be helpful to review the language and laws relating to positive whole number indices.

Terminology

In the expression a^m the number represented by a is called the *base* of the expression; m is the *index*, or the *power* to which the base is raised. (The plural of index is indices).

Multiplication

Multiplying 3^6 by 3^4 gives

$$3^6 \times 3^4 = (3 \times 3 \times 3 \times 3 \times 3 \times 3) \times (3 \times 3 \times 3 \times 3)$$
$$= 3^{10}$$

Clearly it is not necessary to write down all the 3s like that. All you need to do is to add the powers: $6 + 4 = 10$

and so $\qquad 3^6 \times 3^4 = 3^{6+4} = 3^{10}$.

This can be written in general form as

$$a^m \times a^n = a^{m+n}.$$

Another important multiplication rule arises when a base is successively raised to one power and then another, as for example in $(3^4)^2$.

$$(3^4)^2 = (3^4) \times (3^4)$$
$$= (3 \times 3 \times 3 \times 3) \times (3 \times 3 \times 3 \times 3)$$
$$= 3^8$$

In this case the powers to which 3 is raised are multiplied: $4 \times 2 = 8$. Written in general form this becomes:

$$(a^m)^n = a^{mn}$$

Division

In the same way, dividing 3^6 by 3^4 gives

$$3^6 \div 3^4 = \frac{(3 \times 3 \times 3 \times 3 \times 3 \times 3)}{(3 \times 3 \times 3 \times 3)}$$
$$= 3^2$$

In this case you subtract the powers: $6 - 4 = 2$

and so $\qquad 3^6 \div 3^4 = 3^{6-4} = 3^2$.

This can be written in general form as

$$a^m \div a^n = a^{m-n}.$$

Extending the use of indices

Using these rules you can now go on to give meanings to indices which are not positive whole numbers.

Index 0

If you divide 3^4 by 3^4 the answer is 1, since this is a number divided by itself. However you can also carry out the division using the rules of indices to get

$$3^4 \div 3^4 = 3^{4-4}$$
$$= 3^0$$

and so it follows that $\qquad 3^0 = 1.$

The same argument applies to 5^0, 2.9^0 or any other (non-zero) number raised to the power 0; they are all equal to 1.

In general $\qquad a^0 = 1.$

Negative indices

Dividing 3^4 by 3^6 gives

$$3^4 \div 3^6 = \frac{3 \times 3 \times 3 \times 3}{3 \times 3 \times 3 \times 3 \times 3 \times 3}$$
$$3^{4-6} = \frac{1}{3 \times 3}$$

and so $\qquad 3^{-2} = \frac{1}{3^2}.$

This can be generalised to $\qquad a^{-m} = \frac{1}{a^m}$

Fractional indices

What number multiplied by itself gives the answer 3? The answer, as you will know, is the square root of 3, usually written √3. Suppose instead that the square root of 3 is written 3^p; what then is the value of p?

Since $\qquad 3^p \times 3^p = 3^1$

it follows that $\qquad p + p = 1$

and so $\qquad p = \frac{1}{2}$

In other words, the square root of a number can be written as that number raised to the power $\frac{1}{2}$: $\sqrt{a} = a^{1/2}$.

The same argument may be extended to other roots, so that the cube root of a number may be written as that number raised to the power $\frac{1}{3}$, the fourth root corresponds to power $\frac{1}{4}$ and so on:

$$\sqrt[n]{a} = a^{1/n}.$$

In the example of Jupiter's moons, or indeed the moons or planets of any system, the relationship between T and r is of the form

$$T = kr^{1\frac{1}{2}}.$$

Squaring both sides gives

$$T \times T = kr^{1\frac{1}{2}} \times kr^{1\frac{1}{2}}$$

which may be written as $\qquad T^2 = cr^3 \qquad$ (where the constant $c = k^2$).

This is one of Kepler's Laws of Planetary Motion, first stated in 1619.

The use of indices not only allows certain expressions to be written more simply, but also, and this is more important, it makes it possible to carry out arithmetical and algebraic operations (like multiplication and division) on them. These processes are shown in the following examples.

EXAMPLE Write the following numbers as the base 5 raised to a power.

(i) 625   (ii) 1   (iii) $\dfrac{1}{125}$ (iv) $5\sqrt{5}$

Solution

Notice that $5^1 = 5$, $5^2 = 25$, $5^3 = 125$, $5^4 = 625$, ...

(i) $625 = 5^4$ (ii) $1 = 5^0$

(iii) $\dfrac{1}{125} = \dfrac{1}{5^3}$ (iv) $5\sqrt{5} = 5^1 \times 5^{1/2}$

$\qquad\qquad\quad = 5^{-3}$ $= 5^{1\frac{1}{2}}$

EXAMPLE Simplify   (i) $(2^3)^4$ (ii) $27^{1/3}$ (iii) $4^{-2\frac{1}{2}}$

Solution

(i) $(2^3)^4 = 2^{12}$ (ii) $27^{1/3} = \sqrt[3]{27}$ (iii) $4^{-2\frac{1}{2}} = (2^2)^{-\frac{5}{2}}$

$\qquad\qquad = 4096$ $= 3$ $= 2^{2 \times (-\frac{5}{2})}$

$\qquad\qquad\qquad\qquad\qquad\qquad\qquad\qquad\qquad\qquad\qquad = 2^{-5}$

$\qquad\qquad\qquad\qquad\qquad\qquad\qquad\qquad\qquad\qquad\qquad = \dfrac{1}{32}$

EXAMPLE Simplify $8^{\frac{2}{3}}$

Solution

There are two ways to approach this:

(i) $8^{\frac{2}{3}} = (8^2)^{\frac{1}{3}}$ (ii) $8^{\frac{2}{3}} = (8^{\frac{1}{3}})^2$

$\qquad\qquad = (64)^{\frac{1}{3}}$ $= (2)^2$

$\qquad\qquad = 4$ $= 4$

They give the same answer, and both are correct.

EXAMPLE Simplify $\left(4\sqrt{2} \times \dfrac{1}{16} \times \sqrt[5]{32}\right)^2$

Solution

$$\left(4\sqrt{2} \times \frac{1}{16} \times \sqrt[5]{32}\right)^2 = (2^2 \times 2^{\frac{1}{2}} \times 2^{-4} \times (2^5)^{\frac{1}{5}})^2$$

$$= (2^{2\frac{1}{2}} \times 2^{-4} \times 2^1)^2$$

$$= (2^{2\frac{1}{2}-4+1})^2$$

$$= (2^{-\frac{1}{2}})^2$$

$$= 2^{-1}$$

$$= \tfrac{1}{2}$$

Notice how in this example all the terms are expressed as a power of the base 2. When using the laws of indices you should always make sure that all the terms you are multiplying or dividing are written to the same base.

Mixed bases

If a problem involves a combination of different bases, you may well need to split them, using the rule

$$(a \times b)^n = a^n \times b^n$$

before going on to work with each base in turn.

For example, $6^3 = (2 \times 3)^3$

$$= 2^3 \times 3^3$$

which is another way of saying $216 = 8 \times 27$.

Exercise 1A

Answer these questions without using your calculator, but do use it to check your answers.

1. Write the following numbers as powers of 3.

    (a) 81   (b) 1   (c) 27   (d) $\dfrac{1}{27}$ (e) $\sqrt{3}$ (f) 3

2. Write the following numbers as powers of 4.

    (a) 16   (b) 2   (c) $\dfrac{1}{4}$ (d) $\dfrac{1}{2}$ (e) 8   (f) $\dfrac{1}{8}$

3. Write the following numbers as powers of 10.

    (a) 1000   (b) 0.0001   (c) $\dfrac{1}{1000}$ (d) $\sqrt{1000}$ (e) $\dfrac{1}{\sqrt{10}}$

    (f) One millionth

4. Write the following as whole numbers or fractions.

    (a) 2^{-5} (b) $27^{\frac{1}{3}}$ (c) $27^{-\frac{1}{3}}$ (d) $625^{\frac{1}{4}}$ (e) $625^{-\frac{1}{4}}$ (f) 19^{0}

5. Simplify the following.

    (a) $3\sqrt{3} \times 3^{\frac{1}{2}}$ (b) 1000×10^{-2} (c) $\sqrt{25} \times 5^{-2}$ (d) $7^{\frac{1}{2}} \times 7^{-\frac{1}{2}}$

    (e) $\sqrt{(5^{4} \times 5^{6})}$ (f) $\sqrt[3]{(243 \times 81)}$

6. Simplify the following.

    (a) $\dfrac{7^{-2}}{7^{-4}}$ (b) $(3^{\frac{1}{2}})^{2}$ (c) $(3^{2})^{\frac{1}{2}}$ (d) $(\sqrt{27})^{4}$ (e) $\sqrt[4]{(16^{3})}$

    (f) $(4^{-\frac{1}{2}})^{-2}$

7. Simplify

    (a) $64^{\frac{1}{2}} - 64^{\frac{1}{3}}$ (b) $(11^{2})^{3} - (11^{3})^{2}$ (c) $25^{\frac{1}{2}} - \dfrac{5^{-4}}{5^{-5}}$

    (d) $2^{0} + 2^{-1} + 2^{-2} + 2^{-3} + 2^{-4}$ (e) $(3^{\frac{1}{2}})^{2} - (3^{2})^{\frac{1}{2}}$ (f) $3^{1} - 3^{-1}$

Working with square roots

You often meet square roots while doing mathematics, and there are times when it is helpful to be able to manipulate them, rather than just find their values from your calculator. This ensures that you are working with the exact value of the square root, rather than a rounded-off version. It may also keep your work tidier, without long numbers appearing at every line. Sometimes you find that the square roots cancel out anyway, and this may indicate a relationship that you would otherwise have missed.

A number which is partly rational and partly square root, like $(\frac{1}{2} + \sqrt{5})$, is called a *surd*.

The basic fact about the square root of a number is that when multiplied by itself, it gives the number: $\quad \sqrt{5} \times \sqrt{5} = 5$

This can be rearranged to give

$$\sqrt{5} = \frac{5}{\sqrt{5}} \quad \text{and} \quad \frac{1}{\sqrt{5}} = \frac{\sqrt{5}}{5} .$$

Both of these are useful results.

Surds may be added or subtracted just like algebraic expressions, keeping the rational numbers and the square roots separate.

EXAMPLE　Add　$(2 + 3\sqrt{5})$　to　$(3 - \sqrt{5})$

Solution

$$2 + 3\sqrt{5} + 3 - \sqrt{5} = 2 + 3 + 3\sqrt{5} - \sqrt{5}$$
$$= 5 + 2\sqrt{5}$$

EXAMPLE　Simplify　$4(\sqrt{7} - 2\sqrt{2}) + 3(2\sqrt{7} + \sqrt{2})$

Solution

$$4(\sqrt{7} - 2\sqrt{2}) + 3(2\sqrt{7} + \sqrt{2}) = 4\sqrt{7} - 8\sqrt{2} + 6\sqrt{7} + 3\sqrt{2}$$
$$= 10\sqrt{7} - 5\sqrt{2}$$

When multiplying two surds, do so term by term, as in the following examples.

EXAMPLE　Simplify　$(2 + \sqrt{3})^2$

Solution

$$(2 + \sqrt{3})^2 = (2 + \sqrt{3}) \times (2 + \sqrt{3})$$
$$= 2 \times 2 + 2 \times \sqrt{3} + \sqrt{3} \times 2 + \sqrt{3} \times \sqrt{3}$$
$$= 4 + 2\sqrt{3} + 2\sqrt{3} + 3$$
$$= 7 + 4\sqrt{3}$$

EXAMPLE　Simplify　$(\sqrt{2} + \sqrt{3})(\sqrt{2} + 2\sqrt{3})$

Solution

$$(\sqrt{2} + \sqrt{3})(\sqrt{2} + 2\sqrt{3}) = \sqrt{2} \times \sqrt{2} + \sqrt{3} \times \sqrt{2} + \sqrt{2} \times 2\sqrt{3} + \sqrt{3} \times 2\sqrt{3}$$
$$= 2 + \sqrt{6} + 2\sqrt{6} + 2 \times 3$$
$$= 8 + 3\sqrt{6}$$

Notice that in the last example we used the fact that $\sqrt{2} \times \sqrt{3} = \sqrt{(2 \times 3)} = \sqrt{6}$.

EXAMPLE

Simplify $(2 + \sqrt{3}) \times (2 - \sqrt{3})$

Solution

$$(2 + \sqrt{3}) \times (2 - \sqrt{3}) = 2 \times 2 - 2 \times \sqrt{3} + \sqrt{3} \times 2 - \sqrt{3} \times \sqrt{3}$$
$$= 4 - 2\sqrt{3} + 2\sqrt{3} - 3$$
$$= 1$$

You will notice that in the last example the terms involving square roots disappear, leaving an answer that is a rational number. This is because the two expressions to be multiplied together are the factors of the difference of two squares:

$$(a + b)(a - b) = a^2 - b^2.$$

In this case $a = 2$ and $b = \sqrt{3}$, so that

$$(2 + \sqrt{3})(2 - \sqrt{3}) = 2^2 - (\sqrt{3})^2$$
$$= 4 - 3$$
$$= 1$$

This is the basis of a useful technique for simplifying a fraction whose bottom line is a surd. The technique called rationalising the denominator is illustrated in the next example.

EXAMPLE

Simplify by rationalising the denominator $\dfrac{(4+\sqrt{5})}{(3+\sqrt{5})}$

Solution

Multiply both the top line and the bottom line by $(3 - \sqrt{5})$. You can see that this will make the bottom line into an expression of form $(a - b)(a + b)$, which is $(a^2 - b^2)$, and is a rational number.

$$\frac{(4+\sqrt{5})}{(3+\sqrt{5})} = \frac{(4+\sqrt{5}) \times (3-\sqrt{5})}{(3+\sqrt{5}) \times (3-\sqrt{5})}$$
$$= \frac{12-4\sqrt{5}+3\sqrt{5}-5}{9-3\sqrt{5}+3\sqrt{5}-5}$$
$$= \frac{7-\sqrt{5}}{4}$$
$$= 1\tfrac{3}{4} - \tfrac{1}{4}\sqrt{5}$$

NOTE

This technique involves multiplying the top line and the bottom line of a fraction by the same factor. This does not change the value of the fraction: it is equivalent to multiplying the fraction by 1.

Exercise 1B

Leave all answers in this exercise in surd form, where appropriate.

1. Simplify

(a) $\dfrac{7}{\sqrt{7}}$ (b) $\dfrac{9}{3\sqrt{3}}$ (c) $\dfrac{(2+\sqrt{2})}{\sqrt{2}}$ (d) $\dfrac{5\sqrt{5}}{\sqrt{125}}$ (e) $\dfrac{6}{\sqrt{3}}-\dfrac{8}{2\sqrt{2}}$

2. Simplify

(a) $(3+\sqrt{2})+(5+2\sqrt{2})$ (b) $(5+3\sqrt{2})-(5-2\sqrt{2})$

(c) $4(1+\sqrt{5})+3(5+4\sqrt{5})$ (d) $3(\sqrt{2}-\sqrt{5})+5(\sqrt{2}+\sqrt{5})$

(e) $3(\sqrt{2}-1)+3(\sqrt{2}+1)$

3. Simplify

(a) $(3+\sqrt{2})(4-\sqrt{2})$ (b) $(2-\sqrt{3})^2$ (c) $(3-\sqrt{5})(\sqrt{5}+5)$

(d) $(1+\sqrt{5})(1-\sqrt{5})$ (e) $(\frac{1}{2}+\frac{1}{2}\sqrt{3})^2$ (f) $(2-\sqrt{7})(\sqrt{7}-1)$

(g) $(4+\sqrt{2})(4-\sqrt{2})$ (h) $(5+\sqrt{30})(6-\sqrt{30})$ (i) $(\sqrt{7}+\sqrt{5})(\sqrt{7}-\sqrt{5})$

4. Simplify by rationalising the denominator

(a) $\dfrac{(3+\sqrt{2})}{(4-\sqrt{2})}$ (b) $\dfrac{(2-\sqrt{3})}{(1+\sqrt{3})}$ (c) $\dfrac{(3-\sqrt{5})}{(\sqrt{5}+5)}$ (d) $\dfrac{1}{(1-\sqrt{5})}$

(e) $\dfrac{\sqrt{3}}{\left(\frac{1}{2}-\frac{1}{2}\sqrt{3}\right)}$ (f) $\dfrac{2}{(\sqrt{7}-1)}$ (g) $\dfrac{(4+\sqrt{2})}{(4-\sqrt{2})}$ (h) $\dfrac{(6+\sqrt{30})}{(6-\sqrt{30})}$

Logarithms

You can think of multiplication in two ways. Look, for example, at 81×243, which is $3^4 \times 3^5$. You can work out the product using the numbers or you can work it out by adding the powers of a common base – in this case base 3.

Multiplying the numbers: $81 \times 243 = 19683$

Adding the powers of the base 3: $4+5=9: \quad 3^9 = 19683$

Another name for a power is a *logarithm*. Since $81 = 3^4$, you can say that the logarithm to the base 3 of 81 is 4. The word logarithm is often abbreviated to log and that statement would be written:

$$\log_3 81 = 4$$

Notice that since $3^4 = 81$, $3^{\log_3 81} = 81$. This is an example of a general result, that $a^{\log_a x} = x$.

EXAMPLE

(i) Express as logarithms to the base 2 the numbers:

     (a) 64   (b) $\frac{1}{2}$ (c) 1   (d) $\sqrt{2}$

(ii) Show that $2^{\log_2 64} = 64$

Solution

(i) (a) $64 = 2^6$ and so $\log_2 64 = 6$

(b) $\frac{1}{2} = 2^{-1}$ and so $\log_2 \frac{1}{2} = -1$

(c) $1 = 2^0$ and so $\log_2 1 = 0$

(d) $\sqrt{2} = 2^{\frac{1}{2}}$ and so $\log_2 \sqrt{2} = \frac{1}{2}$

(ii) $2^{\log_2 64} = 2^6 = 64$ as required

Logarithms to the base 10

Any positive number can be expressed as a power of 10. Before the days of calculators, logarithms to the base 10 were used extensively as an aid to calculation. There is no need for that nowadays but the logarithm function remains an important part of mathematics, particularly the natural logarithm which you will meet in chapter 5. Base 10 logarithms continue to be a standard feature on calculators, and occur in some specialised contexts: the pH value of a liquid, for example, is a measure of its acidity or alkalinity and is given by $\log_{10}(1/\text{the concentration of H}^+ \text{ ions})$.

Since $\quad 1000 = 10^3, \qquad \log_{10} 1000 = 3.$

Similarly $\qquad\qquad\qquad \log_{10} 100 = 2$

$$\log_{10} 10 = 1$$

$$\log_{10} 1 = 0$$

$$\log_{10} \frac{1}{10} = \log_{10}(10^{-1}) = -1$$

$$\log_{10} \frac{1}{100} = \log_{10}(10^{-2}) = -2$$

and so on.

Investigation

There are several everyday situations in which quantities are measured on logarithmic scales.

What are the relationships between

(i) an earthquake of intensity 7 on the Richter Scale, and one of intensity 8?

(ii) the frequency of the musical note middle C and that of the C above it?

(iii) the intensity of an 85 dB noise level and one of 86 dB?

The laws of logarithms

The laws of logarithms follow from those for indices, derived at the start of this chapter.

Multiplication
Writing $mn = m \times n$ in the form of powers (or logarithms) to the base a and using the result that $x = a^{\log_a x}$ gives

$$a^{\log_a mn} = a^{\log_a m} \times a^{\log_a n}$$

and so $\qquad a^{\log_a mn} = a^{\log_a m + \log_a n}.$

Consequently $\qquad \log_a mn = \log_a m + \log_a n.$

Division
Similarly $\qquad \log_a \left(\dfrac{m}{n} \right) = \log_a m - \log_a n.$

Power 0
Since $\qquad a^0 = 1, \log_a 1 = 0.$

However, it is more usual to state such laws without reference to the base of the logarithms except where necessary, and this convention is adopted in the Key Points summary at the end of the chapter. As well as the laws given above, others may be derived from them, as follows.

Indices

Since $\qquad m^r = m \times m \times m \times \ldots \times m \qquad$ (r times)

it follows that $\qquad \log m^r = \log m + \log m + \log m + \ldots \log m \qquad$ (r times),

and so $\qquad \log m^r = r \log m.$

This result is also true for non-integer values of r and is particularly useful because it allows you to solve equations in which the unknown quantity is the power, as in the next example.

EXAMPLE

Solve the equation $\qquad 2^n = 1000.$

Solution

$$2^n = 1000.$$

Taking logarithms to the base 10 of both sides (since these can be found on a calculator),

$$\log_{10}(2^n) = \log_{10} 1000$$

$$n\log_{10} 2 = \log_{10} 1000$$

$$n = \frac{\log_{10} 1000}{\log_{10} 2}$$

$$= 9.97 \text{ to 3 significant figures.}$$

Notice that most calculators just have 'log' and not 'log$_{10}$' on their keys.

Roots

A similar line of reasoning leads to the conclusion that

$$\log \sqrt[r]{m} = \frac{1}{r}\log m$$

The logic runs as follows:

Since $\qquad \sqrt[r]{m} \times \sqrt[r]{m} \times \sqrt[r]{m} \times \ldots \times \sqrt[r]{m} = m$

it follows that $\qquad\qquad\qquad r\log\sqrt[r]{m} = \log m$

and so $\qquad\qquad\qquad\qquad \log\sqrt[r]{m} = \frac{1}{r}\log m$

The logarithm of a number to its own base

Since $5^1 = 5$, it follows that $\log_5 5 = 1$.

Clearly the same is true for any number, and in general,

$$\log_a a = 1.$$

You will use this result in Chapter 5 to find the value of the number e, the base of natural logarithms.

Reciprocals

Another useful result is that, for any base,

$$\log\left(\frac{1}{n}\right) = -\log n.$$

This is a direct consequence of the division law

$$\log_a\left(\frac{m}{n}\right) = \log_a m - \log_a n$$

with m set equal to 1:

$$\log\left(\frac{1}{n}\right) = \log 1 - \log n$$
$$= 0 - \log n$$
$$= -\log n.$$

If the number n is greater than 1, it follows that $\frac{1}{n}$ lies between 0 and 1 and $\log\left(\frac{1}{n}\right)$ is negative. So for any base (>1), the logarithm of a number between 0 and 1 is negative. You saw an example of this on page 11:

$\log_{10}\left(\frac{1}{10}\right) = -1.$

Activity

Draw the graph of $y = \log_2 x$, taking values of x like $\frac{1}{8}$, $\frac{1}{4}$, $\frac{1}{2}$, 1, 2, 4, 8, 16.
Use your graph to estimate the value of $\sqrt{2}$.

The result $\log\left(\dfrac{1}{n}\right) = -\log n$ is often useful in simplifying expressions within logarithms.

Graphs of logarithms

Whatever the value, a, of the base (>1), the graph of $y = \log_a x$ has the same general shape (shown in figure 1.1).

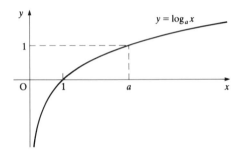

Figure 1.1

- It crosses the x axis at $(1,0)$.

- The curve only exists for positive values of x.

- The line $x = 0$ is an asymptote, and for values of x between 0 and 1 the curve lies below the x axis.

- There is no limit to the height of the curve for large values of x, but its gradient progressively decreases.

- The curve passes through the point $(a, 1)$.

For Discussion

Each of these points can be justified by work that you have already covered. How?

Exponential functions

The relationship $y = \log_a x$ may be rewritten as $x = a^y$, and so the graph of $x = a^y$ is exactly the same as that of $y = \log_a x$. Interchanging x and y has the effect of reflecting the graph in the line $y = x$, and changing the relationship into $y = a^x$, as shown in figure 1.2.

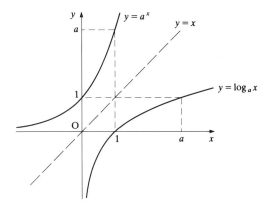

Figure 1.2

The function $y = a^x$ is called an *exponential function*. It is the *inverse* of the logarithm function. Inverse functions are covered in detail in Chapter 3, but, to put it briefly, the effect of applying a function followed by its inverse is to bring you back to where you started.

Thus $\log_a (a^x) = x$, and $a^{(\log_a x)} = x$.

Exercise 1C

1. Write down the values of the following without using a calculator. Use your calculator to check your answers for those questions which use base 10.

   (a) $\log_{10} 10\,000$ (b) $\log_{10}\left(\dfrac{1}{10\,000}\right)$ (c) $\log_{10} \sqrt{10}$ (d) $\log_{10} 1$

   (e) $\log_3 81$ (f) $\log_3\left(\dfrac{1}{81}\right)$ (g) $\log_3 \sqrt{27}$ (h) $\log_3 \sqrt[4]{3}$

   (i) $\log_4 2$ (j) $\log_5\left(\dfrac{1}{125}\right)$

2. Write the following expressions in the form $\log x$ where x is a number.

   (a) $\log 5 + \log 2$ (b) $\log 6 - \log 3$ (c) $2\log 6$

   (d) $-\log 7$ (e) $\tfrac{1}{2}\log 9$ (f) $\tfrac{1}{4}\log 16 + \log 2$

   (g) $\log 5 + 3\log 2 - \log 10$ (h) $\log 12 - 2\log 2 - \log 9$

   (i) $\tfrac{1}{2}\log \sqrt{16} + 2\log (\tfrac{1}{2})$ (j) $2\log 4 + \log 9 - \tfrac{1}{2}\log 144$

3. Use logarithms to the base 10 to solve the following equations.

   (a) $2^n = 1\,000\,000$ (b) $2^n = 0.001$ (c) $1.08^n = 2$
   (d) $1.1^n = 100$ (e) $0.99^n = 0.000\,001$

Exercise 1C continued

4. In *A Man of Property* (the first book in the Forsyte Saga by John Galsworthy), Soames Forsyte states how many years it takes to double your money at 7% compound interest. Given that interest is paid yearly, at the end of the year, what number of years does he say?

Modelling curves

When you obtain experimental data, you are often hoping to establish a mathematical relationship between the variables in question. Should the data fall on a straight line, you can do this easily because you know that a straight line with gradient m and intercept c has equation $y = mx + c$.

EXAMPLE

In an experiment the temperature θ (in °C) was measured at different times t (in seconds), in the early stages of a chemical reaction. The results are shown in the table below.

t	20	40	60	80	100	120
θ	16.3	20.4	24.2	28.5	32.0	36.3

(i)   Plot a graph of θ against t.
(ii)   What is the relationship between θ and t?

Solution

(i)

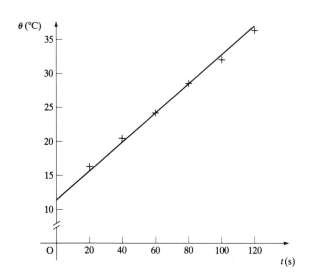

(ii) The graph shows the points lying on a reasonably good straight line and so it is possible to estimate its gradient and intercept.

Intercept: $c = 12.3$

Gradient: $m = \dfrac{36.3 - 16.3}{120 - 20}$

$= 0.2$

In this case the equation is not $y = mx + c$ but $\theta = mt + c$, and so is given by

$$\theta = 0.2t + 12.3.$$

It is often the case, however, that your results do not end up lying on a straight line but on a curve, so that this straightforward technique cannot be applied. The appropriate use of logarithms can convert some curved graphs into straight lines. This is the case if the relationship has one of two forms, $y = kx^n$ or $y = ka^x$.

The techniques used in these two cases are illustrated in the following examples. In theory, logarithms to any base may be used, but in practice you would only use those available on your calculator: logarithms to the base 10, and natural logarithms. The base of natural logarithms is a number denoted by e, 2.71828... In chapter 5 you will see how this apparently unnatural number arises naturally; for the moment what is important is that you can apply the techniques using logarithms to either base. The two examples are worked in base 10. Make sure that you do the subsequent activities, reworking them using natural logarithms.

The natural logarithm of x may be written either as $\log_e x$ or as $\ln x$.

Relationships of the form $y = kx^n$

EXAMPLE

A water pipe is going to be laid between two points and an investigation is carried out as to how, for a given pressure difference, the rate of flow R in litres per second varies with the diameter of the pipe d measured in cm. The following data are collected.

d	1	2	3	5	10
R	0.02	0.32	1.62	12.53	199.80

It is suspected that the relationship between R and d may be of the form

$$R = kd^n$$

where k is a constant.

(i) Explain how a graph of $\log d$ against $\log R$ tells you whether this is a good model for the relationship.

(ii) Make out a table of values of $\log_{10} d$ against $\log_{10} R$ and plot these on a graph.

(iii) If appropriate, use your graph to estimate the values of n and k.

Solution

(i) If the relationship is of the form $R = kd^n$, then taking logarithms gives

$$\log R = \log k + \log d^n$$

or $\qquad\qquad\qquad \log R = n\log d + \log k.$

This equation is of the form $y = mx + c$, with $\log R$ replacing y, n replacing m, $\log d$ replacing x and $\log k$ replacing c. It is thus the equation of a straight line.

Consequently if the graph of $\log R$ against $\log d$ is a straight line, the model $R = kd^n$ is appropriate for the relationship and n is given by the gradient of the graph. The value of k is found from the intercept, $\log k$, of the graph with the vertical axis. If you are working with logarithms to the base 10,

$$\log_{10} k = \text{intercept} \qquad \Rightarrow \qquad k = 10^{\text{intercept}}.$$

If you are working with natural logarithms,

$$\log_e k = \text{intercept} \qquad \Rightarrow \qquad k = e^{\text{intercept}}.$$

(ii) Working to 2 decimal places (you would find it hard to draw the graph to greater accuracy) the logarithmic data are as follows.

$\log_{10} d$	0	0.30	0.48	0.70	1.00
$\log_{10} R$	-1.70	-0.49	0.21	1.10	2.30

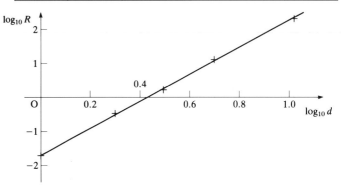

In this case the graph is indeed a straight line with gradient 4 and intercept -1.70, so $n = 4$ and $c = 10^{-1.70} = 0.020$ (to 2 significant figures).

The proposed equation linking R and d is a good model for their relationship, and may be written as

$$R = 0.02d^4$$

Activity

Obtain the same relationship working with natural logarithms.

Exponential relationships

EXAMPLE

The temperature, θ in °C, of a cup of coffee at time t minutes after it is made is recorded as follows.

t	2	4	6	8	10	12
θ	81	70	61	52	45	38

(i) Plot the graph of θ against t.

(ii) Show how it is possible, by drawing a suitable graph, to test whether the relationship between θ and t is of the form $\theta = ka^t$, where k and a are constants.

(iii) Carry out the procedure.

Solution

(i)

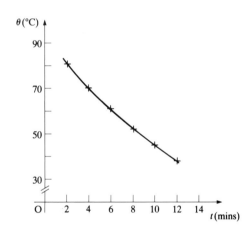

(ii) If the relationship is of the form $\theta = ka^t$, taking logarithms of both sides gives

$$\log \theta = \log k + \log a^t,$$

or $$\log \theta = t\log a + \log k.$$

This is an equation of the form $y = mx + c$ with $\log \theta$ replacing y, t replacing x, $\log a$ replacing m and $\log k$ replacing c. It is thus the equation of a straight line.

Consequently if the graph of $\log\theta$ against t is a straight line, the model $\theta = ka^t$ is appropriate for the relationship, and $\log a$ is given by the gradient of the graph. The value of a is therefore found as $a = 10^{\text{gradient}}$ if logarithms to the base 10 are being used, and as $a = e^{\text{gradient}}$ if natural logarithms are being used. Similarly, the value of k is found from the intercept, $\log k$, of the line with the vertical axis: $k = 10^{\text{intercept}}$ or $e^{\text{intercept}}$.

(iii) The table gives values of $\log_{10} \theta$ for the given values of t

t	2	4	6	8	10	12
$\log_{10} \theta$	1.908	1.845	1.785	1.716	1.653	1.580

The graph of $\log_{10} \theta$ against t is as follows.

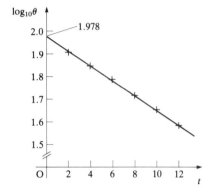

The graph is indeed a straight line so the proposed model is appropriate.

The gradient is -0.029 and so $a = 10^{-0.029} = 0.935$

The intercept is 1.978 and so $k = 10^{1.978} = 95.1$

The relationship between θ and t is given by

$$\theta = 95.1 \times 0.935^t$$

Notice that because the base of the exponential function, 0.935, is less than 1, the function's value decreases rather than increases with t.

Activity
Obtain the same relationship working with natural logarithms.

Exercise 1D

1. The planet Saturn has many moons. The table below gives the mean radius of orbit and the time taken to complete one orbit, for five of the best-known of them.

Moon	Tethys	Dione	Rhea	Titan	Iapetus
Radius R ($\times 10^5$ km)	2.9	3.8	5.3	12.2	35.6
Period T (days)	1.9	2.7	4.5	15.9	79.3

It is believed that the relationship between R and T is of the form $R = kT^n$.

(i)   How can this be tested by plotting $\log T$ against $\log R$?

(ii) Make out a table of values of $\log R$ and $\log T$ and draw the graph.

(iii) Use your graph to estimate the values of k and n.

In 1980 a Voyager spacecraft photographed several previously unknown moons of Saturn. One of these, named 1980 S.27, has a mean orbital radius of $1.4 \times 10^5 \, \text{km}$.

(iv) Estimate how many days it takes this moon to orbit Saturn.

2. The table below shows the area, $A \, \text{cm}^2$, occupied by a patch of mould at time t days since measurements were started.

t	0	1	2	3	4	5
A	0.9	1.3	1.8	2.5	3.5	5.2

It is believed that A may be modelled by a relationship of the form $A = kb^t$.

(i)   Show that the model may be written as $\log A = t\log b + \log k$.

(ii) What graph must be plotted to test this model?

(iii) Plot the graph and use it to estimate the values of b and k.

(iv) Estimate (a) the time when the area of mould was $2 \, \text{cm}^2$; (b) the area of the mould after 3.5 days.

(v) How is this sort of growth pattern described?

3. The inhabitants of an island are worried about the rate of deforestation taking place. A research worker uses records over the last 200 years to estimate the number of trees at different dates. Her results are as follows:

Year	1800	1860	1910	1930	1950	1960	1970	1990
Trees (thousands)	3000	2900	3200	2450	1800	1340	1000	740

(i)   Look at the figures and state when the deforestation may be taken to have begun, and the number of trees at that time.

It is suggested that the number of trees N has been decreasing exponentially with the number of years, t, since that time, so that N may be modelled by the equation.

$$N = ka^t$$

where k and a are constants.

(ii) Explain the meaning of the constant k in this situation, and also why you would expect the value of a to be less than 1.

Exercise 1D continued

(iii) Show that the model may be written as

$$\log N = t\log a + \log k$$

(iv) Plot the graph of $\log N$ against t and use it to estimate the values of k and a.

(v) If this continues to be an appropriate model for the deforestation, in what year can the number of trees be expected to fall below $500\,000$?

4. The time after a train leaves a station is recorded in minutes as t and the distance that it has travelled in metres as s. The results are as follows:

t	0.5	1.0	1.5	2.0	2.5
s	420	1200	2200	3390	4740

It is suggested that the relationship between s and t is of the form $s = kt^n$ where k and n are constants.

(i) Draw the graph of $\log s$ against $\log t$ and explain why this tells you that this model is indeed appropriate.

(ii) Use your graph to estimate the values of k and n.

(iii) Estimate how far the train travelled in its first 100 seconds.

(iv) Explain why you would be wrong to use your results to estimate the distance the train has travelled after 10 minutes.

NOTE: *This model is actually that of a train whose engine is working at constant power, with air resistance ignored.*

5. All but one of the following pairs of readings satisfy, to 3 significant figures, a formula of the type $y = A \times x^B$

x	1.51	2.13	3.50	4.62	5.07	7.21
y	2.09	2.75	4.09	5.10	6.21	7.28

Find the values of A and B, explaining your method.

If the values of x are correct, state which value of y appears to be wrong and estimate what the value should be.

[MEI]

6. The following table gives corresponding values of two variables x and y.

x	3.0	4.3	5.5	8.0	10	11	12
y	2.2	2.6	3.0	3.6	4.0	4.2	4.4

The relation between x and y is known to be of the form $y = kx^n$ where k and n are constants.

By a graphical method, estimate the values of k and n, giving your answers to two significant figures.

[O&C]

7. An experimenter takes observations of a quantity y for various values of a variable x. He wishes to test whether these observations conform to a formula $y = A \times x^B$ and, if so, to find the values of the constants A and B.

Take logarithms of both sides of the formula. Use the result to explain what he should do, what will happen if there is no relationship, and if there is one, how to find A and B.

Carry this out accurately on squared paper for the following observations, and record clearly the resulting formula if there is one.

x	4	7	10	13	20
y	3	3.97	4.74	5.41	6.71

[MEI]

KEY POINTS

- **Indices**

  Multiplication: $\qquad a^m \times a^n = a^{m+n}$

  Division: $\qquad a^m \div a^n = a^{m-n}$

  Power 0: $\qquad a^0 = 1.$

  Negative indices: $\qquad a^{-m} = \dfrac{1}{a^m}$

  Fractional indices: $\qquad a^{1/n} = \sqrt[n]{a}$

  Power of a power: $\qquad (a^m)^n = a^{mn}$

- A function of the form a^x is described as exponential.

- **Logarithms to any base**

  Multiplication: $\qquad \log mn = \log m + \log n.$

  Division: $\qquad \log\left(\dfrac{m}{n}\right) = \log m - \log n.$

  Logarithm of 1: $\qquad \log 1 = 0$

  Powers: $\qquad \log m^r = r\log m$

(continued)

Reciprocals: $\qquad$ $\log\left(\dfrac{1}{n}\right) = -\log n$

Roots: $\qquad$ $\log \sqrt[r]{m} = \dfrac{1}{r}\log m$

Logarithm to its own base: $\quad \log_a a = 1$

- Logarithms may be used to discover the relationship between the variables in two types of situation:

$y = kx^n \qquad$ plot $\log y$ against $\log x$: if a straight line results, n is the gradient, $\log k$ is the intercept;

$y = ka^x \qquad$ plot $\log y$ against x: if a straight line results, $\log a$ is the gradient, $\log k$ is the intercept.

2

Sequences and series

Population, when unchecked, increases in a geometrical ratio. Subsistence increases only in an arithmetical ratio. A slight acquaintance with numbers will show the immensity of the first power in comparison with the second

Thomas Malthus (1798)

For Discussion

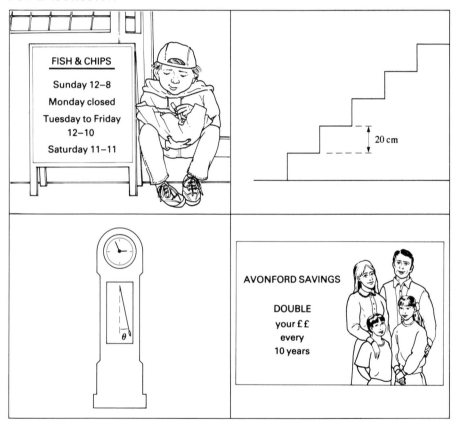

FISH & CHIPS

Sunday 12–8

Monday closed

Tuesday to Friday
12–10

Saturday 11–11

20 cm

AVONFORD SAVINGS

DOUBLE

your £ £

every

10 years

Each of the following sequences is related to one of the pictures above.

(a) 5000, 10 000, 20 000, 40 000, ….

(b) 8, 0, 10, 10, 10, 10, 12, 8, 0, ….

(c) 5, 3.5, 0, −3.5, −5, −3.5, 0, 3.5, 5, 3.5, ….

(d) 20, 40, 60, 80, 100, ….

(i)   Identify which sequence goes with which picture.

(ii)   Give the next few numbers in each sequence.

(iii) Describe the pattern of the numbers in each case.

(iv) Decide whether the sequence will go on for ever, or come to a stop.

Definitions and Notation

A *sequence* is a set of numbers in a given order, like

$$\tfrac{1}{2}, \tfrac{1}{4}, \tfrac{1}{8}, \tfrac{1}{16}, \ldots$$

Each of these numbers is called a *term* of the sequence. When writing the terms of a sequence algebraically, it is usual to denote the position of any term in the sequence by a subscript, so that a general sequence might be written:

$$a_1, a_2, a_3, \ldots, \text{with general term } a_k.$$

For the sequence above, the first term is $a_1 = \tfrac{1}{2}$, the second term is $a_2 = \tfrac{1}{4}$, and so on.

When the terms of a sequence are added together, like

$$\tfrac{1}{2} + \tfrac{1}{4} + \tfrac{1}{8} + \tfrac{1}{16} + \ldots$$

the resulting sum is called a *series*. The process of adding the terms together is called *summation* and indicated by the symbol $\sum$ (the Greek letter sigma), with the position of the first and last terms involved given as *limits*.

So $\quad a_1 + a_2 + a_3 + a_4 + a_5 \quad$ is written $\quad \displaystyle\sum_{k=1}^{k=5} a_k \quad$ or $\quad \displaystyle\sum_{k=1}^{5} a_k.$

In cases like this one, where there is no possibility of confusion, the sum would normally be written more simply as $\quad \displaystyle\sum_{1}^{5} a_k$

If all the terms were to be summed, it would usually be denoted even more simply, as $\displaystyle\sum_{k} a_k$, or even $\displaystyle\sum a_k.$

A sequence may have an infinite number of terms, in which case it is called an *infinite sequence*. The corresponding series is called an *infinite series*.

In mathematics, although the word *series* can describe the sum of the terms of any sequence, it is usually used only when summing the sequence provides some useful or interesting overall result. For example:

$$\pi = 2\sqrt{3}\left[1 + \left(\frac{-1}{3}\right) + 5\left(\frac{-1}{3}\right)^2 + 7\left(\frac{-1}{3}\right)^3 + \ldots\right]$$

$$(1+x)^5 = 1 + 5x + 10x^2 + 10x^3 + 5x^4 + x^5$$

$$\sin x = x - \frac{x^3}{3!} + \frac{x^5}{5!} - \ldots \quad \text{(where } x \text{ is small and measured in radians).}$$

In this book the wording 'sum of a sequence' is often used to mean the sum of the terms of a sequence (i.e. the series).

Patterns in sequences

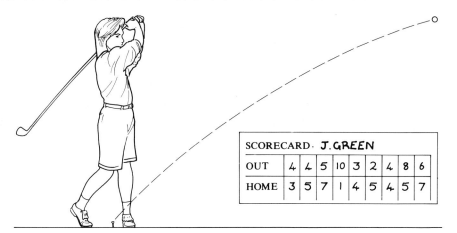

SCORECARD· **J.GREEN**

OUT	4	4	5	10	3	2	4	8	6
HOME	3	5	7	1	4	5	4	5	7

Figure 2.1

Any ordered set of numbers, like the scores of this golfer on an 18 hole round, (figure 2.1) form a sequence. In mathematics, we are particularly interested in those which have a well-defined pattern, often in the form of an algebraic formula linking the terms. The sequences you met at the start of this chapter show various types of pattern.

Arithmetic sequences

A sequence in which the terms increase by the addition of a fixed amount, (or decrease by subtraction of a fixed amount), is described as *arithmetic*. The increase from one term to the next is called the *common difference*.

Thus the sequence $\quad$ 5 $\quad$ 8 $\quad$ 11 $\quad$ 14... $\quad$ is arithmetic with

$$\underbrace{\quad}_{+3}\underbrace{\quad}_{+3}\underbrace{\quad}_{+3}$$

common difference 3.

This sequence can be written algebraically in two quite different ways

(i) $\quad a_k = 2 + 3k$ for $k = 1, 2, 3, \ldots$

$\qquad$ When $k = 1,\qquad a_1 = 2 + 3 = 5$

$\qquad\qquad k = 2,\qquad a_2 = 2 + 6 = 8$

$\qquad\qquad k = 3,\qquad a_3 = 2 + 9 = 11$

> This version has the advantage that the right hand side begins with the first term of the sequence.

$\qquad$ and so on.

$\qquad$ (An equivalent way of writing this is $\qquad a_k = 5 + 3(k - 1)$ for $k = 1, 2, 3 \ldots$)

(ii) $\quad a_1 = 5$

$\qquad a_{k+1} = a_k + 3$ for $k = 1, 2, 3$

Substituting $\quad k = 1 \quad \Rightarrow \quad a_2 = a_1 + 3 = 5 + 3 = 8$

$\qquad\qquad\quad k = 2 \quad \Rightarrow \quad a_3 = a_2 + 3 = 8 + 3 = 11$

$\qquad\qquad\quad k = 3 \quad \Rightarrow \quad a_4 = a_3 + 3 = 11 + 3 = 14$

and so on.

The first form, which defines the value of a_k directly, is called a *deductive definition*. The second, which defines each term by relating it to the previous one, is described as *inductive*. You should be prepared to use either.

EXAMPLE

A sequence is defined deductively by $a_k = 3k - 1$ for $k = 1, 2, 3, \ldots$

(i)   Write down the first six terms of the sequence, and say what kind of sequence it is.

(ii) Find the value of the series $\displaystyle\sum_{k=1}^{k=6} a_k$

Solution

(i)   Substituting $k = 1, 2, \ldots, 6$ in $a_k = 3k - 1$ gives

$a_1 = 3 - 1 = 2 \qquad a_2 = 6 - 1 = 5 \qquad a_3 = 9 - 1 = 8$

$a_4 = 12 - 1 = 11 \qquad a_5 = 15 - 1 = 14 \qquad a_6 = 18 - 1 = 17,$

so the sequence is $\qquad 2, 5, 8, 11, 14, 17, \ldots$

This is an arithmetic sequence with common difference 2.

(ii) $\displaystyle\sum_{k=1}^{k=6} a_k = a_1 + a_2 + a_3 + a_4 + a_5 + a_6$

$\qquad\qquad = \quad 2 + 5 + 8 + 11 + 14 + 17$

$\qquad\qquad = \quad 57$

Arithmetic sequences are dealt with in more detail later in this chapter, on page 33.

Geometric sequences

A sequence in which you find each term by multiplying the previous one by a fixed number is described as *geometric*; the fixed number is the *common ratio*.

Thus $\quad 10 \quad\quad 20 \quad\quad 40 \quad\quad 80\ldots \quad\quad$ is a geometric sequence with common

$\qquad\qquad\qquad \times 2 \quad\quad \times 2 \quad\quad \times 2$

ratio 2.

It may be written algebraically as

$$a_k = 5 \times 2^k \qquad \text{for } k = 1, 2, 3, \ldots \qquad \text{(deductive definition)}$$

or as $\quad a_1 = 10$

$$a_{k+1} = 2a_k \qquad \text{for } k = 1, 2, 3, \ldots \qquad \text{(inductive definition)}$$

Geometric sequences are also dealt with in more detail later on page 39.

Periodic sequences

A sequence which repeats itself at regular intervals is called periodic. In the case of the fish bar at the start of this chapter, the number of hours it is open each day forms the sequence

$a_1 = 8,$	$a_2 = 0,$	$a_3 = 10,$	$a_4 = 10,$	$a_5 = 10,$	$a_6 = 10,$	$a_7 = 12,$
(Sun)	(Mon)	(Tues)	(Wed)	(Thurs)	(Fri)	(Sat)

$a_8 = 8,$	$a_9 = 0 \ldots$
(Sun)	(Mon)

There is no neat algebraic formula for the terms of this sequence but you can see that $\quad a_8 = a_1, \qquad a_9 = a_2, \qquad$ and so on.
In general,

$$a_{k+7} = a_k \text{ for } k = 1, 2, 3, \ldots$$

This sequence is periodic, with period 7. Each term is repeated after 7 terms.

A sequence for which

$$a_{k+p} = a_k \text{ for } k = 1, 2, 3, \ldots \text{ (for a fixed integer, } p)$$

is periodic with period p.

Oscillating sequences

The terms of an oscillating sequence, like

$$5, 6, 5, 4, 5, 6, 5, 4, \ldots$$

lie above and below a middle number, in this case 5 (see figure 2.2).

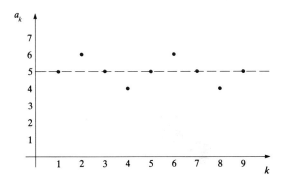

Figure 2.2

In this example, the sequence is also periodic, with period 4. However, some oscillating sequences, like

$$8, -4, 2, -1, \tfrac{1}{2}, \ldots$$

are non-periodic. You will notice that the middle value, zero, is not itself a term in this sequence.

EXAMPLE

A sequence is defined by $a_k = (-1)^k$ for $k = 1, 2, 3, \ldots$

(i)   Write down the first 6 terms of this sequence and describe its pattern.

(ii) Find the value of the series $\displaystyle\sum_{2}^{5} a_k$.

(iii) Describe the sequence defined by $b_k = 5 + (-1)^k \times 2$ for $k = 1, 2, 3, \ldots$

Solution

(i) $a_1 = (-1)^1 = -1$ $a_2 = (-1)^2 = +1$ $a_3 = (-1)^3 = -1$

      $a_4 = (-1)^4 = +1$ $a_5 = (-1)^5 = -1$ $a_6 = (-1)^6 = +1$

The sequence is $-1, +1, -1, +1, -1, +1, \ldots$

It is oscillating, and periodic with period 2. It is also geometric, with common ratio -1.

(ii) $\displaystyle\sum_{2}^{5} a_k = a_2 + a_3 + a_4 + a_5$

           $= (+1) + (-1) + (+1) + (-1)$

           $= 0.$

(iii) $b_1 = 5 + (-1) \times 2 = 3$ $b_2 = 5 + (-1)^2 \times 2 = 7$

       $b_3 = 5 + (-1)^3 \times 2 = 3$ $b_4 = 5 + (-1)^4 \times 2 = 7$

and so on, giving the sequence $3, 7, 3, 7, \ldots$, which is oscillating with period 2.

The device $(-1)^k$ is very useful when writing the terms of oscillating sequences or series algebraically. It is used extensively in mathematics and so the ideas in this example are important.

Sequences with other patterns

There are many other possible patterns in sequences. Figure 2.3 shows a well known children's toy in which squares are stacked to make a tower. The smallest square has sides 1 cm long, and the length of the sides increases in steps of 1 cm.

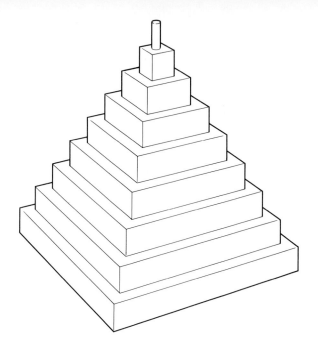

Figure 2.3

The areas of these squares, in cm², form the sequence

$$1^2, 2^2, 3^2, 4^2, 5^2, \ldots. \qquad \text{or} \qquad 1, 4, 9, 16, 25, \ldots.$$

This sequence does not fit any of the patterns described so far. If you subtract each term from the next, however, you will find that the differences form a pattern.

Sequence   1   4   9   16   25   ...

Difference   3   5   7   9   ...

These differences form an arithmetic sequence with common difference 2. The next difference in the sequence is $9 + 2 = 11$, and so the next term in the areas sequence is $25 + 11 = 36$, which is indeed 6^2.

Looking at the differences between the terms often helps you to spot the pattern within a sequence. Sometimes you may need to look at the differences of the differences, or go even further.

Exercise 2A

1. For each of the following sequences, write down the next four terms (assuming the same pattern continues) and describe its pattern as fully as you can.

(a) $7, 10, 13, 16, \ldots$ (b) $8, 7, 6, 5, \ldots$

(c) $10, 30, 90, 270, \ldots$ (d) $64, 32, 16, 8, \ldots$

(e) $2, 2, 2, 5, 2, 2, 2, 5, 2, 2, 2, 5, \ldots$ (f) $4, 6, 7, 6, 4, 2, 1, 2, 4, 6, 7, 6, \ldots$

(g) $1, -2, 4, -8, 16, \ldots$ (h) $2, 4, 6, 8, 2, 4, 6, 8, 2, \ldots$

(i) $4.1, 3.9, 3.7, 3.5, \ldots$ (j) $1, 1.1, 1.21, 1.331, \ldots$

Exercise 2A continued

2. Write down the first 4 terms of the sequences defined below. In each case k takes the values 1, 2, 3, ...

(a) $a_k = 2k + 1$

(b) $a_k = 3 \times 2^k$

(c) $a_k = 2k + 2^k$

(d) $a_k = \frac{1}{k}$

(e) $a_{k+1} = a_k + 3, a_1 = 12$

(f) $a_{k+1} = -a_k, a_1 = -5$

(g) $a_{k+1} = \frac{1}{2} a_k, a_1 = 72$

(h) $a_{k+2} = a_k, a_1 = 4, a_2 = 6$

(i) $a_k = 5 + (-1)^k$

(j) $a_{k+1} = a_k + 2k + 1, a_1 = 1$

3. Find the value of the series $\sum_{1}^{4} a_k$ in each of the following cases.

(a) $a_k = 2 + 5k$

(b) $a_k = 3 \times 2^k$

(c) $a_k = \frac{12}{k}$

(d) $a_{k+1} = 3a_k, a_1 = 1$

(e) $a_k = 2 + (-1)^k$

4. Express each of the following series in the form $\sum_{m}^{n} a_k$, where m and n are integers and a_k is an algebraic expression for the kth term of the series.

(a) $1 + 2 + 3 + \dots + 10$

(b) $21 + 22 + 23 + \dots + 30$

(c) $210 + 220 + 230 + \dots + 300$

(d) $211 + 222 + 233 + \dots + 310$

(e) $190 + 180 + 170 + \dots + 100$

5. Find the values of the following:

(a) $\sum_{1}^{5} k$

(b) $\sum_{1}^{20} k^2 - \sum_{2}^{19} k^2$

(c) $\sum_{0}^{5} (k^2 - 5k)$

6. The Fibonacci sequence is given by

$$1, 1, 2, 3, 5, 8, \dots$$

(i) Write down the sequence of the differences between the terms in this sequence, and comment on what you find.

(ii) Write down the next 3 terms of the Fibonacci sequence.

(iii) Write down the sequence formed by the ratio of one term to the next, $\frac{a_{k+1}}{a_k}$, using decimals. What do you notice about it?

7. The terms of a sequence are defined by

$$a_k = 4 + (-1)^k \times 2$$

(i) Write down the first 6 terms of this sequence.

(ii) Describe the sequence.

(iii) What would be the effect of changing the sequence to

(a) $a_k = 4 + (-2)^k$?

(b) $a_k = 4 + \left(-\frac{1}{2}\right)^k$?

Exercise 2A continued

8. The number π may be calculated using Gregory's series

$$\frac{\pi}{4} = 1 - \frac{1}{3} + \frac{1}{5} - \frac{1}{7} + \ldots$$

(This series was discovered by James Gregory, a Scottish mathematician, in 1671).

(i) Calculate estimates of π using 1, 2, 3, 4, 5, 6, 7 and 8 terms of this series.

(ii) Comment upon your answers to part (i).

9. The harmonic series is given by S in

$$S = 1 + \frac{1}{2} + \frac{1}{3} + \frac{1}{4} + \frac{1}{5} + \frac{1}{6} + \frac{1}{7} + \frac{1}{8} + \ldots$$

(i) Explain why grouping the terms in the form

$$S = 1 + \frac{1}{2} + \left(\frac{1}{3} + \frac{1}{4}\right) + \left(\frac{1}{5} + \frac{1}{6} + \frac{1}{7} + \frac{1}{8}\right) + \left(\ldots\right.$$

allows you to conclude that $\quad S > 1 + \frac{1}{2} + \frac{1}{2} + \frac{1}{2} + \ldots$

(ii) Explain how this method can be used to show that S is infinite.

Investigation

The terms of a sequence may be multiplied together to obtain their product, denoted by Π, (rather than Σ for their sum). Investigate the following products:

(i) $\displaystyle\prod_{k=1}^{n} \frac{k}{k+1} \qquad$ where n is a positive integer greater than 1.

(ii) Wallis' product: $\qquad \frac{\pi}{2} = \left(\frac{2}{1} \times \frac{2}{3}\right) \times \left(\frac{4}{3} \times \frac{4}{5}\right) \times \left(\frac{6}{5} \times \frac{6}{7}\right) \times \ldots$

(This expression was discovered by John Wallis, an English mathematician in 1650.)

Does Wallis' product provide an easier way of finding π than Gregory's series (see question 8)?

Arithmetic sequences and series

Earlier in this chapter, you learned that successive terms of an arithmetic sequence increase (or decrease) by a fixed amount called the common difference, d.

$$a_{k+1} = a_k + d$$

When the terms of an arithmetic sequence are added together, the sum is called an arithmetic series. An alternative name is an *arithmetic progression*, often abbreviated to A.P.

Notation

When describing arithmetic series and sequences in this book, the following conventions will be used:

first term, $a_1 = a$

number of terms $= n$

last term, $a_n = l$

common difference $= d$.

The general term, a_k, is that in position k (i.e. the kth term).

Thus in the arithmetic sequence $\quad$ 5, 7, 9, 11, 13, 15, 17,

$a = 5, \qquad l = 17, \qquad d = 2 \qquad$ and $\qquad n = 7$.

The terms are formed as follows:

$a_1 = a \qquad = 5$

$a_2 = a + d \;\; = 5 + 2 \qquad = 7$

$a_3 = a + 2d = 5 + 2 \times 2 = 9$

$a_4 = a + 3d = 5 + 3 \times 2 = 11$

$a_5 = a + 4d = 5 + 4 \times 2 = 13$

$a_6 = a + 5d = 5 + 5 \times 2 = 15$

$a_7 = a + 6d = 5 + 6 \times 2 = 17$ ◄

> The 7th term is the 1st term (5) plus six times the common difference (2).

You can see that any term is given by the first term plus a number of differences. The number of differences is in each case one less than the number of the term. You can express this mathematically as

$$a_k = a + (k - 1)d.$$

For the last term, this becomes

$$l = a + (n - 1)d$$

These are both general formulae which apply to any arithmetic sequence.

EXAMPLE

Find the 17th term in the arithmetic sequence 12, 9, 6, ...

Solution

In this case $a = 12$ and $d = -3$

Using $\qquad a_k = a + (k - 1)d \qquad$ we obtain

$\qquad\qquad a_{17} = 12 + (17 - 1) \times (-3)$

$\qquad\qquad\qquad = 12 - 48$

$\qquad\qquad\qquad = -36$

The 17th term is -36.

EXAMPLE

How many terms are there in the sequence 11, 15, 19, ..., 643?

Solution

This is an arithmetic sequence with first term, $a = 11$, last term, $l = 643$ and common difference, $d = 4$.

Using the result $\quad\quad l = a + (n - 1)d$, we have

$$643 = 11 + 4\,(n - 1)$$
$$\Rightarrow \quad 4n = 643 - 11 + 4$$
$$\Rightarrow \quad\quad n = 159$$

There are 159 terms

___

Notice that the relationship $\;l = a + (n - 1)d\quad$ may be rearranged to give

$$n = \frac{l - a}{d} + 1.$$

This gives the number of terms in an A.P. directly if you know the first term, the last term and common difference.

The sum of the terms of an arithmetic sequence

When Karl Frederick Gauss (1777–1855) was at school he was always quick to answer mathematics questions. One day his teacher, hoping for half an hour of peace and quiet, told his class to add up all the whole numbers from 1 to 100. Almost at once the 10 year old Gauss announced that he had done it and that the answer was 5050.

Gauss had not of course added the terms, one by one. Instead he wrote the series down twice, once in the given order and once backwards, and added the two together.

$$S = \quad 1 + \quad 2 + \quad 3 + ... + \quad 98 + \quad 99 + 100$$
$$S = 100 + \quad 99 + \quad 98 + ... + \quad 3 + \quad 2 + \quad 1$$

Adding, $2S = 101 + 101 + 101 + ... + 101 + 101 + 101$

Since there are 100 terms in the series,

$$2S = 101 \times 100$$

$$S = 5050$$

The numbers 1, 2, 3, ... , 100 form an arithmetic sequence with common difference 1. Gauss's method can be used for finding the sum of any arithmetic series.

It is common to use the letter S to denote the sum of a sequence. When there is any doubt as to the number of terms that are being summed, this is indicated by a subscript: S_5 indicates 5 terms, S_n indicates n terms.

EXAMPLE Find the value of $8 + 6 + 4 + \ldots + (-32)$

Solution

This is an arithmetic series, with common difference -2. The number of terms, n, may be calculated using

$$n = \frac{l-a}{d} + 1$$

$$n = \frac{-32-8}{-2} + 1$$

$$= 21$$

The sum S of the series is then found as follows:

$$S = \quad 8 + \quad 6 + \ldots - 30 - 32$$
$$S = -32 - 30 - \ldots + \quad 6 + \quad 8$$
$$\overline{2S = -24 - 24 - \ldots - 24 - 24}$$

Since there are 21 terms, this gives $2S = -24 \times 21$, so $S = -12 \times 21 = -252$.

Generalising this method by writing the series in the conventional notation gives:

$$S = \quad\quad [a] \quad\quad + \quad\quad [a+d] \quad\quad + \ldots + [a+(n-2)d] + [a+(n-1)d]$$
$$S = [a+(n-1)d] + [a+(n-2)d] + \ldots + \quad [a+d] \quad + \quad\quad [a]$$
$$\overline{2S = [2a+(n-1)d] + [2a+(n-1)d] + \ldots + [2a+(n-1)d] + [2a+(n-1)d]}$$

Since there are n terms, it follows that

$$S = \tfrac{1}{2}n[2a+(n-1)d]$$

This result may also be written

$$S = \tfrac{1}{2}n(a+l)$$

EXAMPLE Find the sum of the first 100 terms of the sequence

$$1, 1\tfrac{1}{4}, 1\tfrac{1}{2}, 1\tfrac{3}{4}, \ldots$$

Solution

In this arithmetic sequence

$a = 1, d = \tfrac{1}{4}$ and $n = 100$

Using $\quad S = \tfrac{1}{2}n[2a+(n-1)d]$, we have

$$S = \tfrac{1}{2} \times 100\left(2 + 99 \times \tfrac{1}{4}\right)$$

$$= 1337\tfrac{1}{2}$$

<table>
<tr><td>**EXAMPLE**</td><td>Jamila starts a job on salary of £9000 per year, and this increases by an annual increment of £1000. Assuming that apart from the increment, Jamila's salary does not increase, find</td></tr>
</table>

(i)   her salary in the 12th year;

(ii) the length of time she has been working when her total earnings are £100 000.

Solution

Jamila's annual salaries (in pounds) form the arithmetic series

$$9000, \ 10\,000, \ 11\,000, \ ...$$

The first term, $a = 9000$, and the common difference, $d = 1000$.

(i)   Her salary in the 12th year is calculated using

$$a_k \ = a + (k-1)d$$
$$\Rightarrow \quad a_{12} = 9000 + (12-1) \times 1000$$
$$= 20\,000$$

(ii) The number of years that have elapsed when her total earnings are £100 000 is given by

$$S = \tfrac{1}{2}n\big[2a + (n-1)d\big]$$

where $S = 100\ 000$, $a = 9000$ and $d = 1000$.

This gives $100\ 000 = \tfrac{1}{2}n[2 \times 9000 + 1000\,(n-1)]$.

This simplifies to the quadratic equation

$$n^2 + 17n - 200 = 0$$

Factorising,

$$(n-8)(n+25) = 0$$
$$\Rightarrow \quad n = 8 \ \text{ or } \ -25$$

The root $n = -25$ is irrelevant, so the answer is $n = 8$.

Jamila has earned a total of £100 000 after 8 years.

Exercise 2B

1. Are the following sequences arithmetic? If so, state the common difference and the 7th term.

(a) $27, 29, 31, 33, \ ...$ (b) $1, 2, 3, 5, 8, \ ...$ (c) $2, 4, 8, 16, \ ...$

(d) $3, 7, 11, 15, \ ...$ (e) $8, 6, 4, 2, \ ...$

2. The first term of an arithmetic sequence is -8 and the common difference is 3.

(i)   Find the 7th term of the sequence.

(ii) The last term of the sequence is 100. How many terms are there in the sequence?

Exercise 2B continued

3. The first term of an arithmetic sequence is 12, the 7th term is 36 and the last term is 144.
   (i)   Find the common difference.
   (ii) Find how many terms there are in the sequence.

4. There are 20 terms in an arithmetic sequence, whose first term is -5 and last term is 90.
   (i)   Find the common difference.
   (ii) Find the sum of the terms in the sequence.

5. The kth term of an arithmetic sequence is given by
   $$a_k = 14 + 2k$$
   (i)   Write down the first three terms of the sequence.
   (ii) Calculate the sum of the first 12 terms of this sequence.

6. Below is an arithmetic sequence:
   $$120, 114, \ldots, 36.$$
   (i)   How many terms are there in the sequence?
   (ii) What is the sum of the terms in the sequence?

7. The 5th term of an arithmetic sequence is 28 and the 10th term is 58.
   (i)   Find the first term and the common difference.
   (ii) The sum of all the terms in this sequence is 444. How many terms are there?

8. The 6th term of an arithmetic sequence is twice the 3rd term, and the first term is 3. The sequence has 10 terms.
   (i)   Find the common difference.
   (ii) Find the sum of all the terms in the sequence.

9. A set of twelve new stamps is to be issued; the denominations increase in steps of 2p, starting with 1p.
   $$1p, 3p, 5p, 7p, \ldots$$
   (i)   What is the highest denomination of stamp in the set?
   (ii) What is the total cost of the complete set?

10. (i)   Find the sum of all the odd numbers between 50 and 150.
    (ii)   Find the sum of all the even numbers between 50 and 150, inclusive.
    (iii) Find the sum of the terms of the arithmetic sequence with first term 50, common difference 1 and 101 terms.
    (iv)   Explain the relationship between your answers to parts (i), (ii) and (iii).

Exercise 2B continued

11. The first term of an arithmetic sequence is 3000 and the 10th term is 1200.
    (i) Find the sum of the first 20 terms of the sequence.
    (ii) After how many terms does the sum of the sequence become negative?

12. Paul's starting salary in a company is £14 000 and during the time he stays with the company it increases by £500 each year.
    (i) What is his salary in his 6th year?
    (ii) How many years has Paul been working for the company when his total earnings for all his years there are £126 000?

13. A jogger is training for a 10 km charity run. He starts with a run of 400 m; then he increases the distance he runs by 200 m each day.
    (i) How many days does it take the jogger to reach a distance of 10 km in training?
    (ii) What total distance will he have run in training by then?

14. A piece of string 10 m long is to be cut into pieces, so that the lengths of the pieces form an arithmetic sequence.
    (i) The lengths of the longest and shortest pieces are 1 m and 25 cm respectively; how many pieces are there?
    (ii) If the same string had been cut into 20 pieces whose lengths formed an arithmetic sequence, and if the length of the second longest had been 92.5 cm, how long would the shortest piece have been?

Geometric sequences

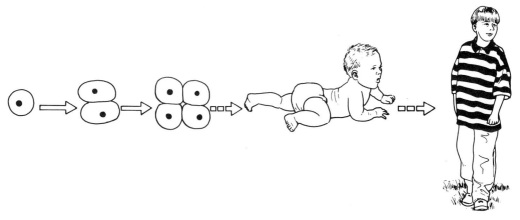

Figure 2.4

A human being begins life as one cell, which divides into two, then four...

The terms of a geometric sequence are formed by multiplying one term by a fixed number, the common ratio, to obtain the next. This can be written inductively as

$$a_{k+1} = ra_k \qquad \text{with first term } a_1.$$

The sum of the terms of a geometric sequence is called a *geometric series*. An alternative name is a *geometric progression*, shortened to G.P.

Notation

When describing geometric sequences in this book, the following conventions will be used.

First term $a_1 = a$

Common ratio $= r$

Number of terms $= n$

The general term a_k is that in position k, (i.e. the kth term).

In the geometric sequence   3, 6, 12, 24, 48, $a = 3, r = 2$ and $n = 5$.

The terms of this sequence are formed as follows:

$$a_1 = a \qquad = 3$$
$$a_2 = a \times r = 3 \times 2 = 6$$
$$a_3 = a \times r^2 = 3 \times 2^2 = 12$$
$$a_4 = a \times r^3 = 3 \times 2^3 = 24$$
$$a_5 = a \times r^4 = 3 \times 2^4 = 48$$

You will see that in each case the power of r is one less than the number of the term: $a_5 = ar^4$ and 4 is one less than 5. This can be written deductively as

$$a_k = ar^{k-1},$$

and the last term is

$$a_n = ar^{n-1}.$$

These are both general formulae which apply to any geometric sequence.

Given two consecutive terms of a geometric sequence, you can always find the common ratio by dividing the later term by the earlier. For example, the geometric sequence .... 5, 8, ... has common ratio $r = \frac{8}{5}$.

EXAMPLE

Find the 7th term in the geometric sequence   8, 24, 72, 216, ...

Solution

In the sequence, the first term, $a = 8$ and the common ratio, $r = 3$.

The kth term of a geometric sequence is given by $a_k = ar^{k-1}$,

and so
$$a_7 = 8 \times 3^6$$
$$= 5832$$

EXAMPLE

How many terms are there in the geometric sequence $0.2, 1, 5, \ldots,$ 390 625?

Solution

The last (nth) term of a geometric sequence is given by $a_n = a \times r^{n-1}$.

In this case $a = 0.2$ and $r = 5$, so

$$390\,625 = 0.2 \times 5^{n-1}$$

$$5^{n-1} = \frac{390\,625}{0.2} = 1\,953\,125$$

Taking logarithms to the base 10 of both sides:

$$(n-1)\log_{10} 5 = \log_{10} 1\,953\,125$$

$$n - 1 = \frac{\log_{10} 1\,953\,125}{\log_{10} 5}$$

$$n - 1 = 9$$

$$n = 10$$

Notice the use of logarithms. This technique, first introduced in Chapter 1, is needed for this sort of problem.

The sum of the terms of a geometric sequence

The origins of Chess are obscure, with several countries claiming the credit for its invention. One story is that it came from China. It is said that its inventor presented the game to the Emperor, who was so impressed that he asked the inventor what he would like as a reward.

'One grain of rice for the first square on the board, 2 for the second, 4 for the third, 8 for the fourth, and so on up to the last square', came the answer.

The Emperor agreed, but it soon became clear that there was not enough rice in the whole of China to give the inventor his reward.

How many grains of rice was the inventor actually asking for?

The answer is the geometric series with 64 terms and common ratio 2:

$$1 + 2 + 4 + 8 + \ldots + 2^{63}.$$

This can be summed as follows.

Call the series S:

$$S = 1 + 2 + 4 + 8 + \ldots + 2^{63}. \qquad \qquad ①$$

Now multiply it by the common ratio, 2:

$$2S = 2 + 4 + 8 + 16 + \ldots + 2^{64} \qquad \qquad ②$$

Then subtract ① from ②

② $2S =$ $2 + 4 + 8 + 16 + ... + 2^{63} + 2^{64}$

① $S =$ $1 + 2 + 4 + 8 +$ $... + 2^{63}$

subtracting: $S = -1 + 0 + 0 + 0 +$ $...$ $+ 2^{64}$

The total number of rice grains requested was therefore $2^{64} - 1$ (which is about 1.85×10^{19}).

For Discussion

How many tonnes of rice is this, and how many tonnes would you expect there to be in China at any time?

(One hundred grains of rice weigh about 2 grammes. The world annual production of all cereals is about 1.8×10^{9} tonnes.)

The method used above can be used to sum any geometric series.

EXAMPLE

Find the value of $0.2 + 1 + 5 + ... + 390\,625$

Solution

This is a geometric series with common ratio 5.

Let $S = 0.2 + 1 + 5 + ... + 390\,625$ ①

Multiplying by the common ratio, 5 gives

$5S = 1 + 5 + 25 + ... + 390\,625 + 1\,953\,125$ ②

Subtracting ① from ②:

$5S =$ $1 + 5 + 25 + ... + 390625 + 1953125$

$S =$ $0.2 + 1 + 5 + 25 + ... + 390625$

$4S = -0.2$ $+ 1953125$

This gives $4S = 1953124.8$

$\Rightarrow$ $S = 488281.2$

The same method can be applied to the general geometric series to give a formula for its value.

$$S = a + ar + ar^2 + ... + ar^{n-1}$$ ①

Multiplying by the common ratio r gives

$$rS = ar + ar^2 + ar^3 + ... + ar^n.$$ ②

Subtracting ① from ②, as before, gives

$$(r-1)S = -a + ar^n$$

$$= a(r^n - 1)$$

so $$S = \frac{a(r^n - 1)}{(r-1)}.$$

The sum to n terms is often written as S_n, and that for an infinite series as S_∞ or S.

For values of r between -1 and 1, the sum to n terms is usually written as

$$S_n = \frac{a(1-r^n)}{(1-r)}.$$

This ensures that you are working with positive numbers in the brackets.

Infinite geometric series

The sequence $1, \frac{1}{2}, \frac{1}{4}, \frac{1}{8}, \frac{1}{16}, \ldots$ is geometric, with common ratio $\frac{1}{2}$.

Summing the terms one by one gives $1, 1\frac{1}{2}, 1\frac{3}{4}, 1\frac{7}{8}, 1\frac{15}{16}, \ldots$

Clearly the more terms you take, the nearer the sum gets to 2. In the limit, as the number of terms tends to infinity, the sum tends to 2.

As $n \to \infty$, $S_n \to 2$.

This is an example of a *convergent* series. The sum to infinity is a finite number.

You can see this by substituting $a = 1$ and $r = \frac{1}{2}$ in the formula for the sum of the series

$$S_n = \frac{a\left(1-r^n\right)}{1-r}$$

giving

$$S_n = \frac{1 \times \left(1-\left(\frac{1}{2}\right)^n\right)}{\left(1-\frac{1}{2}\right)}$$

$$= 2 \times \left(1-\left(\frac{1}{2}\right)^n\right).$$

The larger the number of terms, n, the smaller $\left(\frac{1}{2}\right)^n$ becomes and so the nearer S_n is to the limiting value of 2 (figure 2.5). Notice that $\left(\frac{1}{2}\right)^n$ can never be negative, however large n becomes, so S_n can never exceed 2.

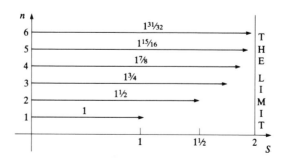

Figure 2.5

In the general geometric sequence $a, ar, ar^2, \ldots$ the terms become progressively smaller in size if the common ratio r is between -1 and 1. This was the case above: r had the value $\frac{1}{2}$. In such cases, the sequence is convergent.

If, on the other hand, the value of r is greater than 1 (or less than -1) the terms in the sequence become larger and larger in size and so it is described as divergent. A sequence corresponding to a value of r of exactly $+1$ consists of the first term a repeated over and over again. A sequence corresponding to a value of r of exactly -1 oscillates between $+a$ and $-a$. Neither of these is convergent. It only makes sense to talk about the sum of an infinite series if it is convergent. Otherwise the sum is undefined.

The condition for a geometric series to converge, $-1 < r < 1$, ensures that as $n \to \infty$, $r^n \to 0$, and so the formula for the sum of a geometric series

$$S_n = \frac{a(1 - r^n)}{(1 - r)}$$

may be re-written for an infinite sequence as

$$S = \frac{a}{1 - r} \quad \text{or} \quad S_\infty = \frac{a}{1 - r}$$

EXAMPLE

Find the sum of the terms of the infinite sequence 0.2, 0.02, 0.002, ...

Solution

This is a geometric sequence with $a = 0.2$ and $r = 0.1$.

Its sum is given by
$$S = \frac{a}{1 - r}$$

$$= \frac{0.2}{1 - 0.1}$$

$$= \frac{0.2}{0.9}$$

$$= \frac{2}{9}$$

NOTE

You may have noticed that the sum of the sequence $0.2 + 0.02 + 0.002 + \ldots$ is $0.22\dot{2}$, and that this recurring decimal is indeed the same as $\frac{2}{9}$.

EXAMPLE

The first three terms of an infinite geometric sequence are 16, 12 and 9.
(i)   Write down the common ratio.
(ii) Find the sum of the terms of the sequence.
(iii) After how many terms is the sum greater than 99.99% of that of the infinite sequence?

Solution

(i) The common ratio is $\frac{3}{4}$.

(ii) The sum of the terms of an infinite geometric series is given by

$$S = \frac{a}{1-r}$$

In this case $a = 16$ and $r = \frac{3}{4}$, so

$$S = \frac{16}{1-\frac{3}{4}} = 64$$

(iii) The sum of a geometric series to n terms, S_n, is given by $\frac{a(1-r^n)}{(1-r)}$.

In this case,

$$S_n = \frac{16\left(1-\left(\frac{3}{4}\right)^n\right)}{\left(1-\left(\frac{3}{4}\right)\right)} = 64\left(1-\left(\frac{3}{4}\right)^n\right)$$

The value of n which gives a sum equal to 99.99% of 64 is given by

$$64\left(1-\left(\frac{3}{4}\right)^n\right) = \frac{99.99}{100}\times 64$$

$$1-\left(\frac{3}{4}\right)^n = 0.9999$$

$$\left(\frac{3}{4}\right)^n = 0.0001$$

Taking logarithms to base 10,

$$n\log_{10}\tfrac{3}{4} = \log_{10}0.0001$$

$$n = \frac{\log_{10}0.0001}{\log_{10}0.75}$$

$$= 32.01$$

So after 33 terms the sum is greater than 99.99% of the sum to infinity of the series.

For Discussion

A paradox

Consider the following arguments.

(a) $S = 1 - 2 + 4 - 8 + 16 - 32 + 64 - \dots$
$\Rightarrow$ $S = 1 - 2(1 - 2 + 4 - 8 + 16 - 32 + \dots)$
        $= 1 - 2S$
$\Rightarrow$ $3S = 1$
$\Rightarrow$ $S = \frac{1}{3}$

(b)
$$S = 1 + (-2 + 4) + (-8 + 16) + (-32 + 64) + \ldots$$
$$\Rightarrow \quad S = 1 + 2 + 8 + 32 + \ldots$$
So S diverges towards $+\infty$.

(c)
$$S = (1 - 2) + (4 - 8) + (16 - 32) + \ldots$$
$$\Rightarrow \quad S = -1 - 4 - 8 - 16 \ldots$$
So S diverges towards $-\infty$.

What is the sum of the series: $\frac{1}{3}$, $+\infty$, $-\infty$, or something else?

Exercise 2C

1. Are the following sequences geometric? If so, state the common ratio and calculate the 7th term.
   (a) $5, 10, 20, 40, \ldots$ (b) $2, 4, 6, 8, \ldots$ (c) $1, -1, 1, -1, \ldots$
   (d) $5, 5, 5, 5, \ldots$ (e) $6, 3, 0, -3, \ldots$ (f) $6, 3, 1\frac{1}{2}, \frac{3}{4}, \ldots$
   (g) $1, 1.1, 1.11, 1.111, \ldots$

2. A geometric sequence has first term 3 and common ratio 2. The sequence has 8 terms.
   (i) Find the last term.
   (ii) Find the sum of the terms in the sequence.

3. The first term of a geometric sequence is 5 and the 5th term is 1280.
   (i) Find the common ratio of the sequence.
   (ii) Find the 8th term of the sequence.

4. A geometric sequence has first term $\frac{1}{9}$ and common ratio 3.
   (i) Find the 5th term.
   (ii) Which is the first term of the sequence which exceeds 1000?

5. (i) Find how many terms there are in the geometric sequence
$$8, 16, \ldots 2048.$$
   (ii) Find the sum of the terms in this sequence.

6. (i) Find how many terms there are in the geometric sequence
$$200, 50, \ldots, 0.195\,312\,5$$
   (ii) Find the sum of the terms in this sequence.

7. The 5th term of a geometric sequence is 48 and the 9th term is 768. All the terms are positive.
   (i) Find the common ratio.
   (ii) Find the first term.
   (iii) Find the sum of the first 10 terms.

8. The first three terms of an infinite geometric sequence are $4, 2$ and 1.
   (i)   State the common ratio of this sequence.
   (ii) Calculate the sum to infinity of its terms.

9. The first three terms of an infinite geometric sequence are $0.7, 0.07,$ 0.007.
   (i)   Write down the common ratio for this sequence.
   (ii) Find, as a fraction, the sum to infinity of the terms of this sequence.
   (iii) Find the sum to infinity of the geometric series $0.7 - 0.07 + 0.007, ...,$ and hence show that $\frac{7}{11} = 0.6\dot{3}$.

10. The first three terms of a geometric sequence are $100, 90$ and 81.
    (i)   Write down the common ratio of the sequence.
    (ii) Which is the position of the first term in the sequence that has value less than 1?
    (iii) Find the sum to infinity of the terms of this sequence.
    (iv) After how many terms is the sum of the sequence greater than 99% of the sum to infinity?

11. A geometric series has first term 4 and its sum to infinity is 5.
    (i)   Find the common ratio.
    (ii) Find the sum to infinity if the first term is excluded from the series.

12. A tank is filled with 20 litres of water. Half the water is removed and replaced with anti-freeze and thoroughly mixed. Half this mixture is then removed and replaced with anti-freeze. The process continues.
    (i)   Find the first five terms in the sequence of amounts of water in the tank at each stage.
    (ii) Find the first five terms in the sequence of amounts of anti-freeze in the tank at each stage.
    (iii) Is either of these sequences geometric? Explain.

13. A pendulum is set swinging. Its first oscillation is through an angle of $30°$, and each succeeding oscillation is through 95% of the angle of the one before it.
    (i)   After how many swings is the angle through which it swings less than $1°$?
    (ii) What is the total angle it has swung through at the end of its 10th oscillation?

14. A ball is thrown vertically upwards from the ground. It rises to a height of $10\,m$ and then falls and bounces. After each bounce it rises vertically to $\frac{2}{3}$ of the height from which it fell.

   (i) Find the height to which the ball bounces after the nth impact with the ground.
   (ii) Find the total distance travelled by the ball from the first throw to the 10th impact with the ground.

15. The first three terms of an arithmetic sequence, a, $a + d$ and $a + 2d$, are the same as the first three terms, a, ar, ar^2, of a geometric sequence. ($a \neq 0$).

Show that this is only possible if $r = 1$ and $d = 0$.

16 . A company offers a ten year contract to an employee. This gives a starting salary of £15 000 a year with an annual increase of 8% of the previous year's salary.
   (i)   Show that the amounts of annual salary form a geometric sequence and write down its common ratio.
   (ii) How much does the employee expect to earn in the tenth year?
   (iii) Show that the total amount earned over the 10 years is nearly £217 500.
   After considering the offer, the employee asks for a different scheme of payment. This has the same starting salary of £15 000 but with a fixed annual pay rise £d.
   (iv) Find d if the total amount paid out over 10 years is to be the same under the two schemes.

[MEI]

Investigations

Mortgage

A householder takes out a 25 year mortgage of £40 000 from a Building Society at a time when the interest rate is 10%. Each year the householder's payments have to cover the interest on the outstanding mortgage, and an element of repayment.

Suppose, as an example, that the householder pays a fixed £5000 a year. In the first year, the interest is 10% of £40 000, i.e. £4000, so £1000 of the £5000 goes towards repayment, leaving £39 000 owing.

In the second year, the interest is 10% of £39 000, i.e. £3900, and so the repayment is £1100, leaving £37 900 owing at the start of the third year, and so on.

If the calculation is done correctly, the final payment at the end of year 25 should repay the last of the mortgage, as well as that year's interest. What is the correct annual payment for such a mortgage? (The figure of £5000 was only a working example).

What is the total amount paid out to the Building Society?

Snowflakes

Draw an equilateral triangle with sides $9\,\text{cm}$ long. Trisect each side and construct equilateral triangles on the middle section of each side as shown in diagram (b). Repeat the procedure for each of the small triangles as shown in (c) and (d) so that you have the first four stages in an infinite sequence.

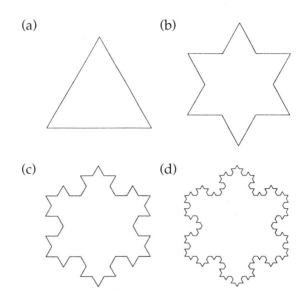

(a)   (b)

(c)   (d)

Calculate the length of the perimeter of the figure for each of the first six steps, starting with the original equilateral triangle.

What happens to the length of the perimeter as the number of steps increases?

Does the area of the figure increase without limit?

Achilles and the Tortoise

Achilles (it is said) once had a race with a tortoise. The tortoise started $100\,\text{m}$ ahead of Achilles and moved at $\frac{1}{10}\,\text{ms}^{-1}$ compared to Achilles' speed of $10\ \text{ms}^{-1}$.

Achilles ran to where the tortoise started only to see that it had moved $1\,\text{m}$ further on. So he ran on to that spot but again the tortoise had moved further on, this time by $0.01\,\text{m}$. This happened again and again: whenever Achilles got to the spot where the tortoise was, it had moved on. Did Achilles ever manage to catch the tortoise?

KEY POINTS

- A sequence is an ordered set of numbers, $a_1, a_2, a_3, \ldots, a_k, \ldots a_n$, where a_k is the general term.

- A series is the sum of the terms of a sequence:

$$a_1 + a_2 + a_3 + \ldots a_n = \sum_{k=1}^{k=n} a_k$$

- In an arithmetic sequence, $\quad a_{k+1} = a_k + d \quad$ where d is a fixed number called the common difference.

- In a geometric sequence, $\quad a_{k+1} = ra_k \quad$ where r is a fixed number called the common ratio.

- In a periodic sequence, $\quad a_{k+p} = a_k \quad$ for a fixed integer, p, called the period.

- In an oscillating sequence the terms rise above and fall below a middle value.

- For an arithmetic sequence with first term a, common difference d and n terms:
  - the kth term, $\quad a_k = a + (k-1)d$;
  - the last term, $\quad l = a + (n-1)d$;
  - the sum of the terms $= \frac{1}{2}n(a + l) = \frac{1}{2}n[2a + (n-1)d]$

- For a geometric sequence with first term a, common ratio r and n terms:
  - the kth term, $a_k = ar^{k-1}$;
  - the last term, $a_n = ar^{n-1}$;
  - the sum of the terms $= \dfrac{a(r^n - 1)}{(r-1)} = \dfrac{a(1-r^n)}{(1-r)}$

- For an infinite geometric series to converge, $-1 < r < 1$. In this case the sum of all the terms is given by $\dfrac{a}{(1-r)}$.

3

Functions

Still glides the stream and shall forever glide;
The form remains, the function never dies.

William Wordsworth

Why fly to Geneva in January?

Several people arriving at Geneva airport from London were asked the main purpose of their visit. Their answers were recorded:

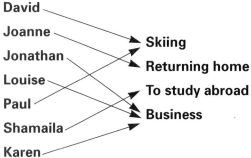

This is an example of a *mapping*.

A mapping is any rule which associates two sets of items. In this example, each of the names on the left is an *object*, or *input*, and each of the reasons on the right is an *image*, or *output*.

For a mapping to make sense or to have any practical application, the inputs and outputs must each form a natural collection or set. The set of possible inputs (in this case, all of the people who flew to Geneva from London in January) is called the *domain* of the mapping. The set of possible outputs (in this case, the set of all possible reasons for flying to Geneva) is called the *co-domain* of the mapping.

The seven people questioned in this example gave a set of four reasons, or outputs. These form the *range* of the mapping for this particular set of inputs. The range of any mapping forms part or all of its co-domain.

Notice that Jonathan, Louise and Karen are all visiting Geneva on business: each person gave only one reason for the trip, but the same reason was given by several people. This mapping is said to be *many-to-one*. A mapping can also be *one-to-one*, *one-to-many*, or *many-to-many*. The relationship between the people and their UK passport numbers will be one-to-one. The relationship between the people and their items of luggage is likely to be one-to-many, and that between the people and the countries they have visited in the last ten years will be many-to-many.

Mathematical mappings

In mathematics, many (but not all) mappings can be expressed using algebra. Here are some examples of mathematical mappings.

Domain: integers	Co-domain: real numbers
Objects	Images

  $-1 \longrightarrow 3$

  $0 \longrightarrow 5$

  $1 \longrightarrow 7$

  $2 \longrightarrow 9$

  $3 \longrightarrow 11$

  General rule: $x \longrightarrow 2x+5$

Domain: integers	Co-domain: real numbers
Objects	Images

  1.9

  2

  2.1

  2.33

  2.52

  3

  2.99

  π

  General rule: Rounded whole numbers   Unrounded numbers

Domain: real numbers	Co-domain: real numbers, y: $-1 \le y \le 1$
Objects	Images

  0

  45

  0

  90

  0.707

  135

  1

  180

  General rule: $x° \longrightarrow \sin x°$

- Domain: quadratic equations   Co-domain: real numbers
  with real roots

  Objects   Images

  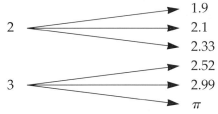

  $x^2 - 4x + 3 = 0$ 0

  $x^2 - x = 0$ 1

  $x^2 - 3x + 2 = 0$ 2

  3

  General rule: $ax^2 + bx + c = 0 \longrightarrow$

  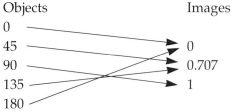

  $$x = \frac{-b - \sqrt{b^2 - 4ac}}{2a}$$

  $$x = \frac{-b + \sqrt{b^2 - 4ac}}{2a}$$

For Discussion

For each of the examples on the previous page,
(i)   decide whether the mapping is one-to-one, many-to-many, one-to-many or many-to-one,
(ii) take a different set of inputs and identify the corresponding range.

Functions

Mappings which are one-to-one or many-to-one are of particular importance, since in these cases there is only one possible image for any object. Mappings of these types are called *functions*. For example, $x \rightarrow x^2$ and $x \rightarrow \cos x°$ are both functions, because in each case for any value of x there is only one possible answer. The mapping of rounded whole numbers onto unrounded numbers is not a function, since, for example, the rounded number 5 could mean any number between 4.5 and 5.5.

There are several different but equivalent ways of writing down a function. For example, the function which maps x onto x^2 can be written in any of the following ways.

- $y = x^2$ ● $f(x) = x^2$ ● $f{:}x \rightarrow x^2$ Read this as 'f maps x onto x²'.

It is often helpful to represent a function graphically, as in the following example, which also illustrates the importance of knowing the domain.

EXAMPLE

Sketch the graph of $y = 3x + 2$ when the domain of x is

(i) $x \in ¥$ (ii) $x \in ¥^+$ (i.e. positive real numbers)   (iii) $x \in ¡$

Solution

(i)   When the domain is $¥$, all values of y are possible. The range is therefore $¥$, also.

(ii) When x is restricted to positive values, all the values of y are greater than 2, so the range is $y > 2$.

(iii) In this case the range is the set of points $\{2, 5, 8, …\}$. These are clearly all of the form $3x + 2$ where x is a natural number, $(0, 1, 2, …)$. This set can be written neatly as $\{3x + 2 : x \in ¡\}$.

The open circle shows that $(0, 2)$ is not part of the line.

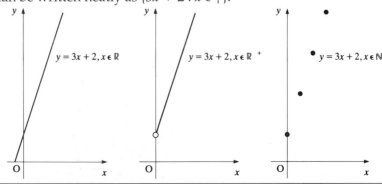

When you draw the graph of a mapping, the x co-ordinate of each point is an input value, the y co-ordinate is the corresponding output value. The table below shows this for the mapping $x \rightarrow x^2$, or $y = x^2$, and figure 3.1 shows the resulting points on a graph.

Input (x)	Output (y)	Point plotted
-2	4	$(-2,4)$
-1	1	$(-1,1)$
0	0	$(0,0)$
1	1	$(1,1)$
2	4	$(2,4)$

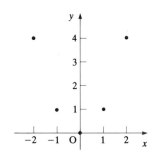

Figure 3.1

If the mapping is a function, there is one and only one value of y for every value of x in the domain. Consequently the graph of a function is a simple curve or line going from left to right, with no doubling back and no breaks or gaps.

Figure 3.2 illustrates some different types of mapping. The graphs in (a) and (b) illustrate functions, those in (c) and (d) do not.

(a) One-to-one

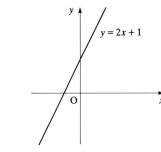

(b) Many-to-one

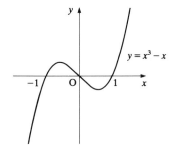

(c) One-to-many

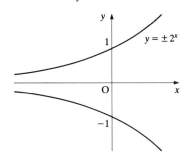

(d) Many-to-many

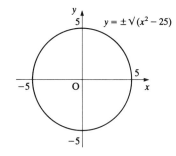

Figure 3.2

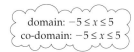

domain: $-5 \le x \le 5$
co-domain: $-5 \le x \le 5$

Exercise 3A

1. Describe each of the following mappings as either one-to-one, many-to-one, one-to-many or many-to-many, and say whether it represents a function. In each case state whether the co-domain and range are equal.

(a)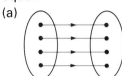

(b)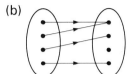

(c)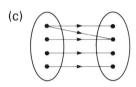

(d)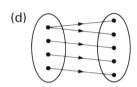

(e)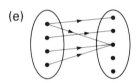

(f)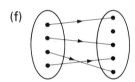

(g)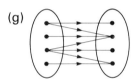

(h)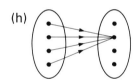

2. For each of the following mappings
   (i)   write down a few examples of inputs and corresponding outputs;
   (ii) state the type of mapping (one-to-one, many-to-one, etc.);
   (iii) suggest suitable domains and co-domains.

   (a) Words $\longrightarrow$ number of letters they contain
   (b) Side of a square $\longrightarrow$ its perimeter
   (c) Natural numbers $\longrightarrow$ the number of factors (including 1 and the number itself)
   (d) $x \longrightarrow 2x - 5$
   (e) $x \longrightarrow \sqrt{x}$
   (f) The volume of a sphere $\longrightarrow$ its radius
   (g) The volume of a cylinder $\longrightarrow$ its height
   (h) The length of a side of a regular hexagon $\longrightarrow$ its area
   (i) $x \longrightarrow x^2$

Exercise 3A continued

3. (a) A function is defined by $f(x) = 2x - 5$. Write down the values of
   (i) $f(0)$ (ii) $f(7)$ (iii) $f(-3)$

   (b) A function is defined by g:(polygons) $\longrightarrow$ (number of sides).
   What are
   (i) g(triangle)   (ii) g(pentagon)   (iii) g(decagon)

   (c) The function t maps Celsius temperatures onto Fahrenheit
   temperatures. It is defined by t: $C \longrightarrow \dfrac{9C}{5} + 32$. Find
   (i) $t(0)$ (ii) $t(28)$ (iii) $t(-10)$ (iv) the value of C when $t(C) = C$

4. Find the range of each of the following functions. (You may find it
   helpful to draw the graph first.)
   (a) $f(x) = 2 - 3x$ $x \geqslant 0$
   (b) $f(\theta) = \sin \theta$ $0° \leq \theta \leq 180°$
   (c) $y = x^2 + 2$ $x \in \{0, 1, 2, 3, 4\}$
   (d) $y = \tan \theta$ $0° < \theta < 90°$
   (e) $f : x \longrightarrow 3x - 5$ $x \in \mathbb{R}$
   (f) $f : x \longrightarrow 2^x$ $x \in \{-1, 0, 1, 2\}$
   (g) $y = \cos x$ $-\frac{\pi}{2} \leq x \leq \frac{\pi}{2}$
   (h) $f : \theta \longrightarrow \sec \theta$ $\theta \in \mathbb{R}$
   (i) $f(x) = \dfrac{1}{1+x^2}$ $x \in \mathbb{R}$
   (j) $f(x) = \sqrt{(x-3)} + 3$ $x \geq 3$

5. The mapping f is defined by $f(x) = x^2$ $0 \leq x \leq 3$
                                                         $f(x) = 3x$ $3 \leq x \leq 10$

   The mapping g is defined by $g(x) = x^2$ $0 \leq x \leq 2$
                                                         $g(x) = 3x$ $2 \leq x \leq 10$

   Explain why f is a function and g is not.

Using transformations to sketch the curves of functions

You already know how to sketch the curves of many functions. Frequently, curve sketching can be made easier by relating the equation of the function to that of a standard function of the same form. This allows you to map the points on the standard curve to equivalent points on the curve you need to draw. You can think of the process as first mapping the function of the standard curve onto the function of the curve you want to draw and then mapping the standard curve (the object) onto the curve you are drawing (the image).

The mappings you will use for curve sketching are called transformations. There are several types of transformation, each with different effects, and you will find that by using them singly or in combination you can sketch a large variety of curves much more quickly.

Translations

Activity

(If possible, use a graphics calculator or computer package for this activity.)

Draw the graphs of $y = x^2$, $y = x^2 + 3$, and $y = x^2 - 2$ on the same axes. What do you notice?

Can you see how the second and third graphs could be obtained from the first one?

How could your findings be generalised?

Repeat the procedure using the graphs of $y = x^2$, $y = (x - 2)^2$ and $y = (x + 3)^2$. What happens now?

Figure 3.3 shows the graphs of $y = x^2$ and $y = x^2 + 3$. For any given value of x, the y co-ordinate for the second curve is 3 units more than the y co-ordinate for the first curve.

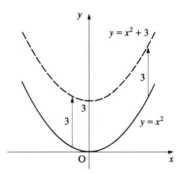

Figure 3.3

Although the curves appear to get closer together as you move away from the line of symmetry, their vertical separation is in fact constant. Each of the vertical arrows is three units long.

You can see that the graphs of $y = x^2 + 3$ and $y = x^2$ are exactly the same shape, but $y = x^2$ has been translated through 3 units in the positive y direction to obtain $y = x^2 + 3$.

Similarly, $y = x^2 - 2$ could be obtained by translating $y = x^2$ through 2 units in the negative y direction (i.e. -2 units in the positive y direction).

In general, for any function $f(x)$, the curve $y = f(x) + s$ can be obtained from that of $y = f(x)$ by translating it through s units in the positive y direction.

What about the relationship between the graphs of $y = x^2$ and $y = (x-2)^2$? Figure 3.4 shows the graphs of these two functions. Again, these curves have exactly the same shape, but as you can see, this time they are separated by a constant 2 units in the x direction.

You may find it surprising that $y = x^2$ moves in the positive x direction when 2 is subtracted from x. It happens because x must be correspondingly larger if $(x - 2)$ is to give the same output that x did in the first mapping.

Notice that the axis of symmetry of the curve $y = (x - 2)^2$ is the line $x = 2$.

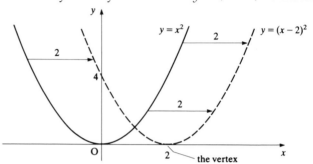

Figure 3.4

In general, the curve with equation $y = f(x - t)$ can be obtained from the curve with equation $y = f(x)$ by a translation of t units in the positive x direction.

Combining these results, $y = f(x - t) + s$ is obtained from $y = f(x)$ by a translation of s units in the positive y direction and t units in the positive x direction. This is called a translation, and is represented by the vector $\begin{pmatrix} t \\ s \end{pmatrix}$.

NOTE

You would usually write $y = f(x - t) + s$ with y as the subject, but this is equivalent to $y - s = f(x - t)$. This form emphasises that subtracting a number from x or y moves the graph in the positive x or y direction.

EXAMPLE

Sketch the curve $y = \sin x$ for $0° \leq x \leq 180°$ and show how it can be used to obtain the graph of $y = \sin(x + 90°)$.

Solution

Re-writing $\sin(x + 90°)$ as $\sin(x - (-90°))$, you can see that the graph of $y = \sin(x + 90°)$ is obtained from the graph of $y = \sin x$ by a translation of $-90°$ in the positive x direction, i.e. $90°$ in the negative x direction.

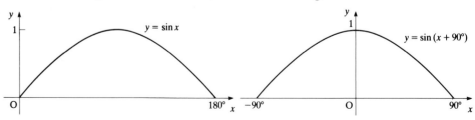

Notice that the resulting graph is the same as that of $y = \cos x$

In Pure Mathematics 1 you met the technique of completing the square of a quadratic function. Using this technique, any quadratic expression of the form $y = x^2 + bx + c$ can be written as $y = (x - t)^2 + s$, so its graph can be sketched by relating it to the graph of $y = x^2$.

EXAMPLE

(i) Find values of s and t such that $x^2 - 2x + 5 \equiv (x - t)^2 + s$.

(ii) Sketch the graph of $y = x^2 - 2x + 5$ and state the position of its vertex and the equation of its axis of symmetry.

Solution

Since the form of the right hand side is given, it is easiest to expand this and then compare coefficients:

$$x^2 - 2x + 5 \equiv (x - t)^2 + s$$
$$\Rightarrow \quad x^2 - 2x + 5 \equiv x^2 - 2xt + t^2 + s$$

Comparing coefficients gives $t = 1$ and $s = 4$.

Re-writing the equation as $y = (x-1)^2 + 4$ shows that the curve can be obtained from the graph of $y = x^2$ by a translation of 1 unit in the positive direction, and 4 units in the positive y direction, i.e. a translation $\begin{pmatrix} 1 \\ 4 \end{pmatrix}$.

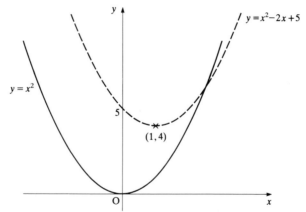

The vertex is (1, 4) and the axis of symmetry is the line $x = 1$.

For Discussion

Would you be justified in finding the values of s and t by substituting, for example, $x = 0$ and $x = 1$ in the identity $x^2 - 2x + 5 \equiv (x - t)^2 + s$?

Activity

Draw the line $y = 2x + 3$. Find the equation obtained when the line is translated through $\begin{pmatrix} 4 \\ 1 \end{pmatrix}$.

(i)   by a graphical method

(ii) by transforming the equation.

EXAMPLE

The diagram shows part of the scalloped bottom of a rollerblind.

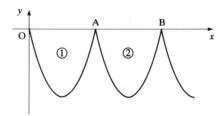

The equation of curve ① with respect to the x and y axes shown is $y = x^2 - 4x$ for $0 \leq x \leq 4$.

Find the equation of curve ②.

Solution

The equation of curve ① can be factorised to give $y = x(x - 4)$, so when $y = 0$, $x = 0$ or 4.

Clearly, O is $(0, 0)$ and A is $(4, 0)$.

Curve ② is therefore obtained from curve ① by a translation of 4 units in the positive x direction.

To find the equation of curve ②, you replace x by $(x - 4)$ in the equation of curve ①.

Using the factorised form, the equation of curve ② is

$$y = (x - 4)[(x - 4) - 4] \qquad \text{for } 0 \leq x - 4 \leq 4$$

$$\Rightarrow \qquad y = (x - 4)(x - 8) \qquad \text{for } 4 \leq x \leq 8$$

One-way stretches

Activity

Draw the graphs of $y = \sin x$ and $y = 2 \sin x$ on the same axes, for $0° \leq x \leq 180°$. If you are using a graphics calculator or computer package you could easily add $y = \frac{1}{2}\sin x$ and $y = 3 \sin x$ to the display.

Describe what happens.

You will have noticed that for any value of x, the y co-ordinate of the point on the curve $y = 2 \sin x$ is exactly double that on the curve $y = \sin x$, as shown in figure 3.5.

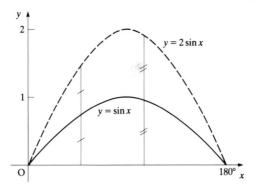

Figure 3.5

This is the equivalent of the curve being stretched parallel to the y axis. Since each y co-ordinate is doubled, this is called a *stretch of scale factor 2 parallel to the y axis.*

NOTE

The equation $y = 2 \sin x$ could also be written as $\dfrac{y}{2} = \sin x$, so dividing y by 2 gives a stretch of scale factor 2 in the y direction.

In general, for any curve $y = f(x)$, and any value of a greater than 0, $y = af(x)$ is obtained from $y = f(x)$ by a stretch of scale factor a parallel to the y axis.

Activity

Now try superimposing the graphs of $y = \sin x$ and $y = \sin 2x$ for $0° \leq x \leq 180°$.

You will see that the graph of $y = \sin 2x$ is compressed parallel to the x axis, so that for any value of y, the x co-ordinate of the point on the curve $y = \sin 2x$ will be exactly half of that on the curve $y = \sin x$ (figure 3.6).

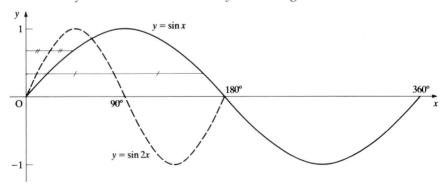

Figure 3.6

Since the curve is compressed to half its original size, this is referred to as a stretch of scale factor $\frac{1}{2}$ parallel to the x axis.

3

In general, for any curve $y = f(x)$ and any value of a greater than 0, $y = f(ax)$ is obtained from $y = f(x)$ by a stretch of scale factor $\frac{1}{a}$ parallel to the x axis. Similarly $y = f\left(\frac{x}{a}\right)$ corresponds to a stretch of scale factor a parallel to the x axis.

NOTE

This is as you would expect: dividing x by a gives a stretch of scale factor a in the x direction, just as dividing y by a gives a stretch of scale factor a in the y direction.

EXAMPLE

Starting with the curve $y = \cos x$, show how transformations can be used to sketch the curves

(i) $y = 2\cos 3x$　　(ii) $y = 3 + \cos\frac{x}{2}$　　(iii) $y = \cos(2x - 60°)$.

Solution

(i) The curve with equation $y = \cos 3x$ is obtained from the curve with equation $y = \cos x$ by a stretch of scale factor $\frac{1}{3}$ parallel to the x axis. There will therefore be one complete oscillation of the curve in 120° (instead of 360°).

The curve of $y = 2\cos 3x$ is obtained from that of $y = \cos 3x$ by a stretch of scale factor 2 parallel to the y axis. The curve therefore oscillates between $y = 2$ and $y = -2$ (instead of $y = 1$ and $y = -1$).

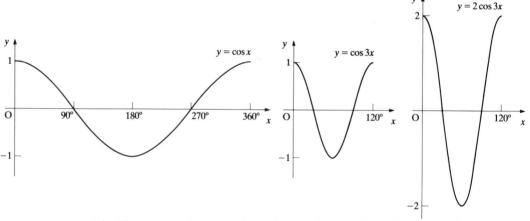

(ii) The curve of $y = \cos\frac{x}{2}$ is obtained from that of $y = \cos x$ by a stretch of scale factor 2 in the x direction. There will therefore be one complete oscillation of the curve in 720° (instead of 360°).

The curve of $y = 3 + \cos\frac{x}{2}$ is obtained from that of $y = \cos\frac{x}{2}$ by a translation $\begin{pmatrix} 0 \\ 3 \end{pmatrix}$.

The curve therefore oscillates between $y = 4$ and $y = 2$.

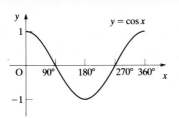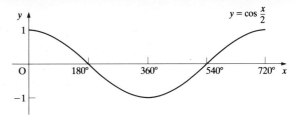

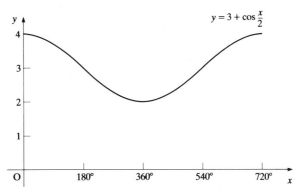

(iii) The curve of $y = \cos(x - 60°)$ is obtained from that of $y = \cos x$ by a translation of $60°$.

The curve of $y = \cos(2x - 60°)$ is obtained from that of $y = \cos(x - 60°)$ by a stretch of scale factor $\frac{1}{2}$ parallel to the x axis.

You must be careful to perform the transformations in the correct order. It is always a good idea to check your results using a graphics calculator.

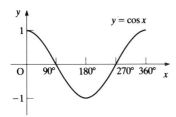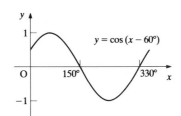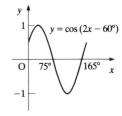

EXAMPLE

(i) Find the values of a, p and q when $y = 2x^2 + 4x - 1$ is written in the form $y = a[x+p)^2 + q]$.

(ii) Show how the graph can be obtained from the graph of $y = x^2$ by successive transformations, and list the transformations in the order in which they are applied.

Solution

Expanding the equivalent expression

$$a[(x+p)^2 + q] \equiv a[x^2 + 2px + p^2 + q]$$
$$\equiv ax^2 + 2apx + a(p^2 + q)$$

Comparing the coefficients in $y = 2x^2 + 4x - 1$ with those above gives

coefficient of x^2: $a = 2$

Functions

coefficient of x: $2ap = 4$, which gives $p = 1$

constant term: $a(p^2+q) = -1$, which gives $q = -1\frac{1}{2}$

The equation of the curve can be written as $y = 2[(x+1)^2 - 1\frac{1}{2}]$

To sketch the graph, start with the curve $y = x^2$.

The curve $y = x^2$ becomes $y = (x+1)^2 - 1\frac{1}{2}$ by applying the translation $\begin{pmatrix} -1 \\ -1\frac{1}{2} \end{pmatrix}$.

The curve $y = (x+1)^2 - 1\frac{1}{2}$ becomes $y = 2[(x+1)^2 - 1\frac{1}{2}]$ by applying a stretch of scale factor 2 parallel to the y axis.

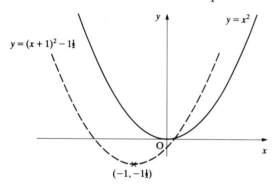

The translation $\begin{pmatrix} -1 \\ -1\frac{1}{2} \end{pmatrix}$

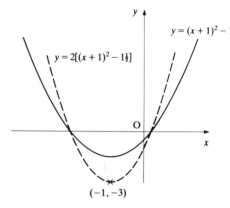

The stretch of scale factor 2 parallel to the y axis

Notice on the graph how the stretch doubles the y co-ordinate of every point on the curve, including the turning point.

Points on the x axis have a zero y co-ordinate, so are unchanged.

Exercise 3B

1. Starting with the graph of $y = x^2$, state transformations which can be used to sketch the following curves. Specify the transformations in the order in which they are used, and where there is more than one stage in the sketching of the curve, state each stage. State the equation of the line of symmetry.

(a) $y = x^2 - 2$

(b) $y = (x + 4)^2$

(c) $y = 4x^2$

(d) $3y = x^2$

(e) $y = (x - 3)^2 - 5$

(f) $y = x^2 - 2x$

(g) $y = x^2 - 4x + 3$

(h) $y = 2x^2 + 4x - 1$

(i) $y = 3x^2 - 6x - 2$

2. Starting with $y = \sin x$, state transformations which can be used to sketch the following curves. Specify the transformations in the order

Exercise 3B continued

in which they are used, and where there is more than one stage in the sketching of the curve, state each stage.

(a) $y = \sin(x - 90°)$ (b) $y = \sin 3x$ (c) $2y = \sin x$

(d) $y = \sin \frac{x}{2}$ (e) $y = 2 + \sin 3x$

3. Starting with $y = \cos x$ state transformations which can be used to sketch the following curves. Specify the transformations in the order in which they are used, and where there is more than one stage in the sketching of the curve, state each stage.

(a) $y = \cos(x + 60°)$ (b) $3y = \cos x$

(c) $y = \cos x + 1$ (d) $y = \cos 2(x + 90°)$

4. For each of the following curves
   (i) sketch the curve;
   (ii) identify the curve as being the same as one of the following:

$$y = \pm \sin x, \qquad y = \pm \cos x, \qquad \text{or} \qquad y = \pm \tan x.$$

(a) $y = \sin(x + 360°)$ (b) $y = \sin(x + 90°)$
(c) $y = \tan(x - 180°)$ (d) $y = \cos(x - 90°)$
(e) $y = \cos(x + 180°)$

5. (i) Show that $x^2 + 6x + 5$ can be written in the form $(x + 3)^2 + a$ where a is a constant to be determined.
   (ii) Sketch the graph of $y = x^2 + 6x + 5$, giving the equation of the axis of symmetry and the co-ordinates of the vertex.

6. Given that $f(x) = x^2 - 6x + 11$, find values of p and q such that $f(x) \equiv (x - p)^2 + q$.
   On the same set of axes, sketch the curves

   (i) $y = f(x)$ and (ii) $y = f(x + 4)$, labelling clearly which is which.

7. The diagram shows the graph of $y = f(x)$ which has a maximum point at $(-2, 2)$, a minimum point at $(2, -2)$, and passes through the origin.

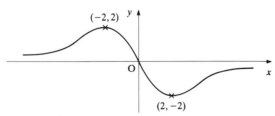

Sketch the following graphs, using a separate set of axes for each graph, and indicating the co-ordinates of the turning points.

(a) $y = 2f(x)$ (b) $y = f(x - 2)$
(c) $y = f(2x)$ (d) $y = 2 + f(x)$
(e) $y = f(x + 2) - 2$ (f) $y = 2f\left(\frac{x}{2}\right)$

Exercise 3B continued

8. A firm can produce a maximum of 50 machines per week, and its income and expenditure are given in pounds by the following equations:

$$\text{total income} = 40x$$

$$\text{total expenditure} = 600 + 0.5x^2$$

where x is the number of machines produced per week.

(i) Sketch both of these graphs on the same axes

(ii) Find the smallest number of machines which must be produced each week to ensure that the firm makes a profit.

Reflections

Activity

Sketch the curves of $y = f(x)$ and $y = -f(x)$ for each of the following functions.

(a) x^2 (b) $\sin x$ (c) $x^3 - 6x^2 + 11x - 6$.

Describe the relationship between the graphs of $y = f(x)$ and $y = -f(x)$ in these cases. Would you be confident to generalise this result?

Figure 3.7 shows the graphs of $y = \cos x$ and $y = -\cos x$ for $0° \leq x \leq 180°$. For any particular value of x, the y co-ordinates of the two graphs have the same magnitude but opposite signs. The graphs are reflections of each other in the x axis.

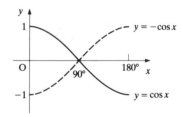

Figure 3.7

In general, starting with the graph of $y = f(x)$ and replacing $f(x)$ by $-f(x)$ gives a reflection in the x axis. This is the equivalent of replacing y by $-y$ in the equation. In the next activity you investigate the effect of replacing x by $-x$.

Activity

Sketch the curves of $y = f(x)$ and $y = f(-x)$ for each of the following functions.

(a) x^2 (b) $\sin x$ (c) $x^3 - 6x^2 + 11x - 6$.

Describe the relationship between the graphs of $y = f(x)$ and $y = f(-x)$ in these cases. Would you be confident to generalise this result?

Figure 3.8 shows the graph of $y = 2x + 1$, a straight line with gradient 2 passing through (0, 1). The graph of $y = 2(-x) + 1$ (which can be written as $y = -2x + 1$) is a straight line with gradient -2, and as you can see it is a reflection of the line $y = 2x + 1$ in the y axis.

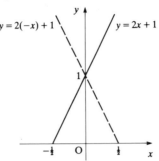

Figure 3.8

In general, starting with the graph of $y = f(x)$ and replacing x by $(-x)$ gives a reflection in the y axis.

EXAMPLE

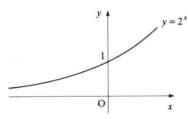

The diagram shows the graph of $y = 2^x$. The curve passes through the point (0, 1).
Sketch, on separate diagrams, the graphs of
(i) $y = 2^{-x}$ (ii) $y = -(2^x)$.

Solution

(i) Replacing x by $-x$ reflects the curve in the y axis.

(ii) The equation $y = -2^x$ can be written as $-y = 2^x$. Replacing y by $-y$ reflects the curve in the x axis.

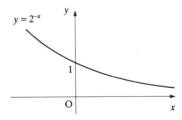

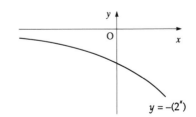

The general quadratic curve

You are now able to relate any quadratic curve to that of $y = x^2$.

EXAMPLE

(i) Write the equation $y = 1 + 4x - x^2$ in the form $y = a[(x + p)^2 + q]$

(ii) Show how the graph of $y = 1 + 4x - x^2$ can be obtained from the graph of $y = x^2$ by a succession of transformations, and list the transformations in the order in which they are applied.

(iii) Sketch the graph.

Solution

(i) If $\qquad 1 + 4x - x^2 \equiv a[(x + p)^2 + q]$

then $\qquad -x^2 + 4x + 1 \equiv ax^2 + 2apx + a(p^2 + q)$

Comparing coefficients of x^2: $\qquad a = -1$.

Comparing coefficients of x: $\qquad 2ap = 4$, $\qquad$ giving $p = -2$.

Comparing constants: $\qquad a(p^2 + q) = 1$, $\qquad$ giving $q = -5$.

The equation is $\qquad y = -[(x - 2)^2 - 5]$.

(ii) The curve $y = x^2$ becomes the curve $y = (x - 2)^2 - 5$ by applying the translation $\begin{pmatrix} 2 \\ -5 \end{pmatrix}$.

The curve $y = (x - 2)^2 - 5$ becomes the curve $y = -[(x-2)^2 - 5]$ by applying a reflection in the x axis.

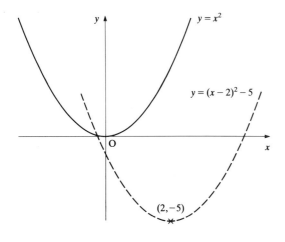

Translation $\begin{pmatrix} 2 \\ -5 \end{pmatrix}$

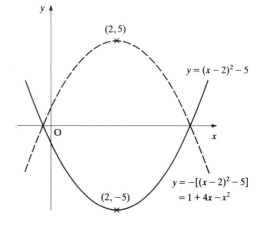

Reflection in x axis

Exercise 3C

1. Starting with the graph of $y = x^2$, state transformations which can be used to sketch the following curves. Specify the transformations in the order in which they are used, and where there is more than one stage in the sketching of the curve, state each stage. State the equation of the line of symmetry.
   (a) $y = -2x^2$ (b) $y = 4 - x^2$ (c) $y = 2x - 1 - x^2$

2. For each of the following curves
   (i) sketch the curve
   (ii) identify the curve as being the same as one of the following:
$$y = \pm \sin x, \qquad y = \pm \cos x, \qquad \text{or} \qquad y = \pm \tan x.$$
   (a) $y = \cos(-x)$ (b) $y = \tan(-x)$
   (c) $y = \sin(180° - x)$ (d) $y = \tan(180° - x)$
   (e) $y = \sin(-x)$

3. (i) Write the expression $x^2 - 6x + 14$ in the form $(x - a)^2 + b$ where a and b are numbers which you are to find.
   (ii) Sketch the curves $y = x^2$ and $y = x^2 - 6x + 14$ and state the transformation which maps $y = x^2$ onto $y = x^2 - 6x + 14$.
   (iii) The curve $y = x^2 - 6x + 14$ is reflected in the x axis. Write down the equation of the image.

4. (i) Sketch the curve with equation $y = x^2$.

   Given that $f(x) = (x-2)^2 + 1$ sketch the curves with the following equations on separate diagrams. Label each curve and give the co-ordinates of its vertex and the equation of its axis of symmetry.
   (ii) $y = f(x)$ (iii) $y = -f(x)$ (iv) $y = f(x+1) + 2$

   [MEI]

5. Write the expression $2x^2 + 4x + 5$ in the form $a(x + b)^2 + c$ where a, b and c are numbers to be found.

   Use your answer to *write down* the co-ordinates of the minimum point on the graph of $y = 2x^2 + 4x + 5$.

   [O & C]

6.

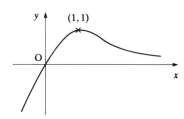

The diagram shows the graph of $y = f(x)$. The curve passes through the origin and has a maximum point at (1, 1).

Exercise 3C continued

Sketch, on separate diagrams, the graphs of

(a) $y = f(x) + 2$ (b) $y = f(x+2)$ (c) $y = f(2x)$,

giving the co-ordinates of the maximum point in each case.

[UCLES]

7. The circle with equation $x^2 + y^2 = 1$ is stretched with scale factor 3 parallel to the x axis and with scale factor 2 parallel to the y axis. Sketch both curves on the same graph, and write down the equation of the new curve. (It is an ellipse).

8. In each of the diagrams below, the curve drawn with a dashed line is obtained as a mapping of the curve $y = f(x)$ using a single transformation. It could be a translation, a one-way stretch or a reflection. In each case, write down the equation of the image (dashed) in terms of $f(x)$.

(a)

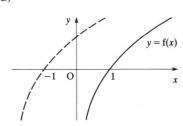

(b)

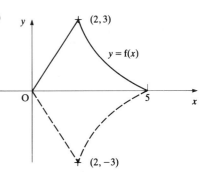

(c)

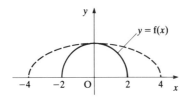

(d)

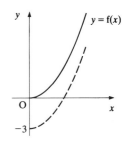

(e)

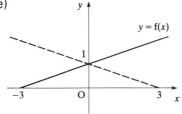

(f)

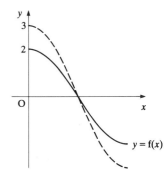

Composite Functions

It is possible to combine functions in several different ways, and you have already met some of these. For example, if $f(x) = x^2$ and $g(x) = 2x$, then you could write

$$f(x) + g(x) = x^2 + 2x.$$

In this example, two functions are added.

Similarly if $f(x) = x$ and $g(x) = \sin x$, then

$$f(x).g(x) = x \sin x.$$

In this example, two functions are multiplied.

Sometimes you need to apply one function and then apply another to the answer. You are then creating a *composite function* or a *function of a function*.

EXAMPLE

A new mother is bathing her baby for the first time. She takes the temperature of the bath water with a thermometer which reads in Celsius, but then has to convert the temperature to degrees Fahrenheit to apply the rule that her own mother taught her:

> At one o five
> He'll cook alive
> But ninety four
> is rather raw.

Write down the two functions that are involved, and apply them to readings of

(i) 30°C   (ii) 38°C   (iii) 45°C

Solution

The first function converts the Celsius temperature C into a Fahrenheit temperature, F:

$$F = \frac{9C}{5} + 32.$$

The second function maps Fahrenheit temperatures onto the state of the bath.

$$F \leq 94 \qquad \text{Too cold}$$
$$95 \leq F \leq 104 \qquad \text{All right}$$
$$F \geq 105 \qquad \text{Too hot}$$

This gives

(i)   30°C $\longrightarrow$ 86°F $\longrightarrow$ too cold

(ii) 38°C $\longrightarrow$ 100.4°F $\longrightarrow$ all right

(iii) 45°C $\longrightarrow$ 113°C $\longrightarrow$ too hot

In this case the composite function would be (to the nearest degree)

$$C \leq 34°C \quad \text{too cold}$$
$$35°C \leq C \leq 40°C \quad \text{all right}$$
$$C \geq 41°C \quad \text{too hot}$$

Read this as 'g of f of x'.

In algebraic terms, a composite function is constructed as

$$\text{Input } x \quad \xrightarrow{\ f\ } \quad \text{Output } f(x)$$

$$\text{Input } f(x) \quad \xrightarrow{\ g\ } \quad \text{Output } g[f(x)] \text{ (or } gf(x)).$$

Thus the composite function $gf(x)$ should be performed from right to left: start with x then apply f and then g.

For example, if f is the rule 'square the input value' and g is the rule 'add 1', then

$$x \quad \xrightarrow[\text{square}]{f} \quad x^2 \quad \xrightarrow[\text{add one}]{g} \quad x^2 + 1$$

So

$$gf(x) = x^2 + 1$$

Notice that $gf(x)$ is not the same as $fg(x)$, since for $fg(x)$ you must apply g first. In the example above, this would give

$$x \quad \xrightarrow[\text{add one}]{g} \quad (x + 1) \quad \xrightarrow[\text{square}]{f} \quad (x + 1)^2$$

and so

$$fg(x) = (x + 1)^2$$

Clearly this is not the same result.

Figure 3.9 illustrates the relationship between the domains and co-domains of the functions f and g, and the co-domain of the composite function gf.

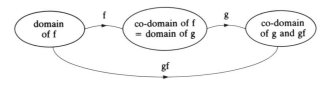

Figure 3.9

EXAMPLE

Given that $f(x) = 2x$, $g(x) = x^2$, and $h(x) = \frac{1}{x}$, find

(i) $fg(x)$ (ii) $gf(x)$ (iii) $gh(x)$
(iv) $f^2(x)$ (v) $fgh(x)$ (vi) $hfg(x)$

Solution

(i) $fg(x) = f[g(x)]$ (ii) $gf(x) = g[f(x)]$

$\qquad\quad = f(x^2)$ $\qquad\qquad = g(2x)$

$\qquad\quad = 2x^2$ $\qquad\qquad = (2x)^2$

$\qquad\qquad\qquad\qquad\qquad\qquad\quad = 4x^2$

(iii) $gh(x) = g[h(x)]$

$$= g\left(\frac{1}{x}\right)$$

$$= \frac{1}{x^2}$$

(iv) $f^2(x) = f[f(x)]$

$$= f(2x)$$
$$= 2(2x)$$
$$= 4x$$

(v) $fgh(x) = f[gh(x)]$

$$= f\left(\frac{1}{x^2}\right) \quad \text{using (iii)}$$

$$= \frac{2}{x^2}$$

(vi) $hfg(x) = h[fg(x)]$

$$= h(2x^2) \quad \text{using (i)}$$

$$= \frac{1}{2x^2}$$

Inverse functions

Look at the mapping $x \rightarrow x + 2$ with domain and co-domain the set of integers.

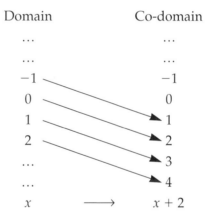

The mapping is clearly a function, since for every input there is one and only one output, the number that is two greater than that input.

This mapping can also be seen in reverse. In that case each number maps onto the number two less than itself: $x \rightarrow x - 2$. The reverse mapping is also a function because for any input there is one and only one output. The reverse mapping is called the *inverse function*, f^{-1}.

Function: $\quad\quad\quad f : x \rightarrow x + 2 \quad\quad x \in \mathbb{Z}$

Inverse function: $\quad f^{-1} : x \rightarrow x - 2 \quad\quad x \in \mathbb{Z}$

For a mapping to be a function which also has an inverse function, every object in the domain must have one and only one image in the co-domain, and vice-versa. This can only be the case if the mapping is one-to-one.

So the condition for a function f to have an inverse function is that, over the given domain and co-domain, f represents a one-to-one mapping. This is a common situation, and many inverse functions are self evident as in the following examples, for all of which the domain and co-domain are the real numbers.

$$f : x \rightarrow x - 1; \qquad f^{-1} : x \rightarrow x + 1$$
$$g : x \rightarrow 2x; \qquad g^{-1} : x \rightarrow \tfrac{1}{2}x$$
$$h : x \rightarrow x^3; \qquad h^{-1} : x \rightarrow \sqrt[3]{x}$$

For Discussion

Some of the following mappings are functions which have inverse functions, and others are not.
(i)   Decide which mappings fall into each category, and for those which do not have inverse functions, explain why.
(ii) For those which have inverse functions, how can the functions and their inverses be written down algebraically?

(a) Temperature measured in Centigrade → temperature measured in Fahrenheit.
(b) Marks in an examination → grade awarded.
(c) Distance measured in light years → distance measured in metres.
(d) Number of stops travelled on the London Underground → fare.

You can decide whether an algebraic mapping is a function, and whether it has an inverse function, by looking at its graph. The curve or line representing a one-to-one mapping does not double back on itself, has no turning points and covers the full domain and co-domain. Figure 3.10 illustrates the functions given above.

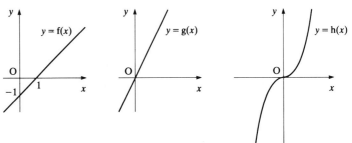

Figure 3.10

Now look at $f(x) = x^2$ for $x \in \mathbb{R}$ (figure 3.11). You can see that there are two distinct input values giving the same output: $f(2) = f(-2) = 4$. When you want to reverse the effect of the function, you need a mapping which for a single input of 4 gives two outputs, -2 and $+2$. Such a mapping is not a function.

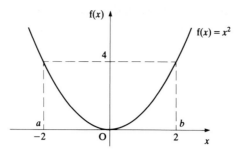

Figure 3.11

If the domain of $f(x) = x^2$ is restricted to $\mathbb{R}^+$ (the set of positive real numbers), you have the situation shown in figure 3.12. This shows that the function which is now defined is one-to-one. The inverse function is given by $f^{-1}(x) = \sqrt{x}$, since the sign $\sqrt{}$ means 'the positive square root of'.

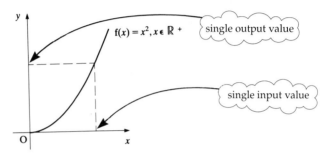

Figure 3.12

It is often helpful to restrict the domain of a function so that its inverse is also a function. When you use the inv sin (or arcsin) key on your calculator the answer is restricted to the range $-90°$ to $90°$, and is described as the *principal value*. Although there are infinitely many roots of the equation $\sin x = 0.5$ ($\ldots -330°, -210°, 30°, 150°, \ldots$), only one of these, $30°$, lies in the restricted range and this is the value your calculator will give you.

The graph of a function and its inverse

Activity

For each of the following functions, work out the inverse function, and draw the graphs of both the original and the inverse on the same axes, using the same scale on both axes.

(i) $f(x) = x^2$ $x \in \mathbb{R}^+$ (ii) $f(x) = 2x$ (iii) $f(x) = x + 2$ (iv) $f(x) = x^3 + 2$

Look at your graphs and see if there is any pattern emerging.

Try out a few more functions of your own to check your ideas.

Make a conjecture about the relationship between the graph of a function and its inverse.

You have probably realised by now that the graph of the inverse function is the same shape as that of the function, but reflected in the line $y = x$. To see why this is so, think of a function $f(x)$ mapping a onto b; (a,b) is clearly a point on the graph of $f(x)$. The inverse function $f^{-1}(x)$, maps b onto a and so (b,a) is a point on the graph of $f^{-1}(x)$.

The point (b,a) is the reflection of the point (a,b) in the line $y = x$. This is shown for a number of points in figure 3.13.

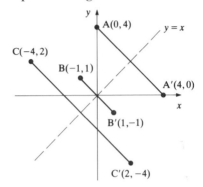

Figure 3.13

This result can be used to obtain a sketch of the inverse function without having to find its equation, provided that the sketch of the original function uses the same scale on both axes.

Finding the algebraic form of the inverse function

To find the algebraic form of the inverse of a function $f(x)$, you should start by changing notation and writing it in the form $y =\ldots$

Since the graph of the inverse function is the reflection of the graph of the original function in the line $y = x$, it follows that you may find its equation by interchanging y and x in the equation of the original function. You will then need to make y the subject of your new equation. This procedure is illustrated in the next example.

EXAMPLE

Find $f^{-1}(x)$ when $f(x) = 2x + 1$

Solution

The function $f(x)$ is given by $\qquad y = 2x + 1$

Interchanging x and y gives $\qquad x = 2y + 1$

Re-arranging to make y the subject: $\quad y = \dfrac{x-1}{2}$

So $\qquad f^{-1}(x) = \dfrac{x-1}{2}$

Sometimes the domain of the function f will not include the whole of $\mathbb{R}$. When any real numbers are excluded from the domain of f, it follows that they will be excluded from the co-domain of f^{-1}, and vice-versa (see figure 3.14).

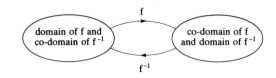

Figure 3.14

EXAMPLE   Find $f^{-1}(x)$ when $f(x) = 2x - 3$ and the domain of f is $x \geq 4$.

Solution

	Domain	Co-domain
Function: $y = 2x - 3$	$x \geq 4$	$y \geq 5$
Inverse function: $x = 2y - 3$	$x \geq 5$	$y \geq 4$

Rearranging the inverse function to make y the subject,

$$y = \frac{x+3}{2}.$$

The full definition of the inverse function is therefore

$$f^{-1}(x) = \frac{x+3}{2} \qquad \text{for } x \geq 5.$$

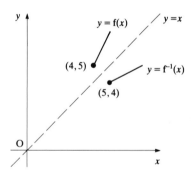

You can see that the inverse function is the reflection of a restricted part of the line $y = f(x)$.

EXAMPLE   (i) Find $f^{-1}(x)$ when $f(x) = x^2 + 2$, $x \geq 0$.

(ii) Find (a) $f(7)$ and (b) $f^{-1} f(7)$. What do you notice?

Solution

(i)

	Domain	Co-domain
Function: $y = x^2 + 2$	$x \geq 0$	$y \geq 2$
Inverse function: $x = y^2 + 2$	$x \geq 2$	$y \geq 0$

Rearranging the inverse function to make y its subject:

$$y^2 = x - 2$$

This gives $y = \pm\sqrt{x-2}$, but since we know the co-domain of the inverse function to be $y \geq 0$ we can write

$$y = +\sqrt{x-2} \quad \text{or just} \quad \sqrt{x-2}.$$

The full definition of the inverse function is therefore

$$f^{-1}(x) = \sqrt{x-2} \qquad \text{for } x \geq 2.$$

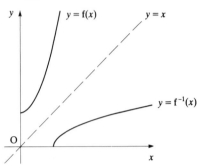

(ii) (a) $f(7) = 7^2 + 2 = 51$

(b) $f^{-1} f(7) = \sqrt{(51-2)} = 7$

Notice that applying the function followed by its inverse has brought us back to the original input value.

Part (ii) of the last example illustrates an important general result. For any function $f(x)$ with an inverse $f^{-1}(x)$, $f^{-1}f(x) = x$. Similarly $ff^{-1}(x) = x$. The effects of a function and its inverse can be thought of as cancelling each other out.

EXAMPLE Find the inverse of the function $f(x) = 10^x$, and sketch $f(x)$ and $f^{-1}(x)$ on the same diagram.

Solution

The function $f(x)$ is given by $y = 10^x$.

Interchanging x and y, the inverse function is given by

$$x = 10^y$$

This can be written as $\log_{10} x = y$, so the inverse function is

$$f^{-1}(x) = \log_{10} x$$

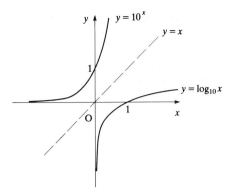

NOTE   *You may recall a similar diagram from p15 in Chapter 1.*

For Discussion

Many calculators have a function and its inverse on the same key, for example log and 10^x, $\sqrt{}$ and x^2, sin and arcsin, ln and e^x.

(i)   With some calculators you can enter a number, apply x^2 and then $\sqrt{}$, and come out with a slightly different number. How is this possible?

(ii)   Explain what happens if you find sin 199° and then the arcsin of the answer.

Inverse trigonometrical functions

The functions sine, cosine and tangent are all many-to-one mappings, so their inverse mappings are one-to-many. Thus the problem 'Find sin 30°' has only one solution, 0.5, whilst 'Find θ such that sin $\theta = 0.5$' has infinitely many solutions. You can see this from the graph of $y = \sin \theta$ (figure 3.15).

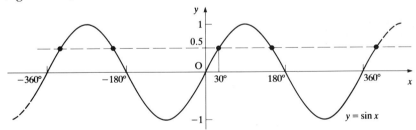

Figure 3.15

In order to define inverse functions for sine, cosine and tangent, a restriction has to be placed on the domain of each so that it becomes a one-to-one mapping.

The restriction of the domain determines the principal values for that trigonometrical function. The restricted domains are not all the same. They are listed below.

Function	Domain (degrees)	Domain (radians)
$y = \sin \theta$	$-90° \leq \theta \leq 90°$	$-\frac{\pi}{2} \leq \theta \leq \frac{\pi}{2}$
$y = \cos \theta$	$0° \leq \theta \leq 180°$	$0 \leq \theta \leq \pi$
$y = \tan \theta$	$-90° < \theta < 90°$	$-\frac{\pi}{2} < \theta < \frac{\pi}{2}$

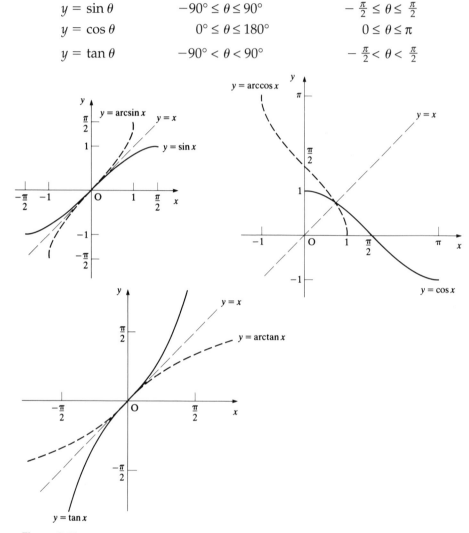

Figure 3.16

Figure 3.16 shows the graph of each trigonometrical function over its restricted domain, and that of its corresponding inverse function. The inverse functions have been drawn using the reflection property, and since this requires that the same scale is used on both axes, the angle must be plotted in radians rather than degrees.

Exercise 3D

1. The functions f, g and h are defined by $f(x) = x^3$, $g(x) = 2x$ and $h(x) = x+2$. Find each of the following, in terms of x.

   (a) fg   (b) gf   (c) fh   (d) hf   (e) fgh

   (f) ghf   (g) g^2 (h) $(fh)^2$ (i) h^2

2. Find the inverses of the following functions

   (a) $f(x) = 2x + 7$ (b) $f(x) = 4 - x$

   (c) $f(x) = \dfrac{4}{2 - x}$ (d) $f(x) = x^2 - 3$ $x \geqslant 0$

3. The function f is defined by $f(x) = (x - 2)^2 + 3$ for $x \geq 2$.

   (i) Sketch the graph of $f(x)$.

   (ii) On the same axes, sketch the graph of $f^{-1}(x)$ without finding its equation.

4. Express the following in terms of the functions f: $x \to \sqrt{x}$ and g: $x \to x+4$.

   (i) $x \to \sqrt{(x + 4)}$ (ii) $x \to x+8$

   (iii) $x \to \sqrt{(x + 8)}$ (iv) $x \to \sqrt{x} + 4$

5. The functions f, g and h are defined by

   $$f(x) = \frac{3}{x - 4} \qquad g(x) = x^2 \qquad h(x) = \sqrt{(2 - x)}$$

   (i) For each function, state any real values of x for which it is not defined.

   (ii) find the inverse functions f^{-1} and h^{-1}.

   (iii) Explain why g^{-1} does not exist when the domain of g is $\mathbb{R}$.

   (iv) Suggest a suitable domain for g so that g^{-1} does exist.

   (v) Is the domain for the composite function fg the same as for the composite function gf? Give reasons for your answer.

6. A function f is defined by

   $$f: x \to \tfrac{1}{x} \qquad x \in \mathbb{R}, x \neq 0, x \neq 1.$$

   Find (i) ff(x)   (ii) fff(x)   (iii) $f^{-1}(x)$ (iv) $f^{999}(x)$

7. The function f is defined by

   $$f: x \to 4x^3 + 3 \qquad x \in \mathbb{R}.$$

   Give the corresponding definition of f^{-1}.

   State the relationship between the graphs of f and f^{-1}.

   [UCLES]

Exercise 3D continued

8. (i) Show that $x^2 + 4x + 7 = (x + 2)^2 + a$, where a is to be determined.

   (ii) Sketch the graph of $y = x^2 + 4x + 7$, giving the equation of its axis of symmetry and the co-ordinates of its vertex.

   The function f is defined by $f : x \to x^2 + 4x + 7$ and has as its domain the set of all real numbers.

   (iii) Find the range of f.

   (iv) Explain, with reference to your sketch, why f has no inverse with its given domain. Suggest a domain for f for which it has an inverse.

   [MEI]

Even, odd and periodic functions

Several of the curves with which you are familiar have symmetry of one form or another. For example

- the curve of any quadratic in x has a line of symmetry parallel to the y axis;

- the curve of $y = \cos x$ has the y axis as a line of symmetry;

- the curves of $y = \sin x$ and $y = \tan x$ have rotational symmetry of order 2 about the origin.

- all the trigonometrical graphs have a repeating pattern (translational symmetry).

In this section you will be looking at particular types of symmetry.

Even functions

A function is *even* if its graph has the y axis as a line of symmetry. This is true for all three of the functions in figure 3.17.

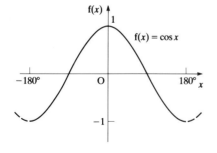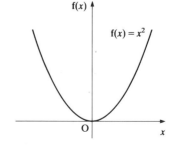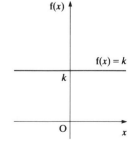

Figure 3.17

Reflecting a curve $y = f(x)$ in the y axis gives the curve $y = f(-x)$, so a curve which has the y axis as a line of symmetry satisfies the condition

$$f(-x) = f(x).$$

This relationship can be used to check whether a function is even, without drawing its graph.

EXAMPLE Show that the function $f(x) = x^4 - 2x^2 + 3$ is an even function.

Solution

$$\begin{aligned}
f(-x) &= (-x)^4 - 2(-x)^2 + 3 \\
&= x^4 - 2x^2 + 3 \\
&= f(x),
\end{aligned}$$

so the function is even.

In general, if $f(x)$ is any polynomial function containing only even powers of x, then $f(x)$ is an even function.

Odd functions

A function whose curve has rotational symmetry of order 2 about the origin, like the curves shown in figure 3.18, is called an *odd function*.

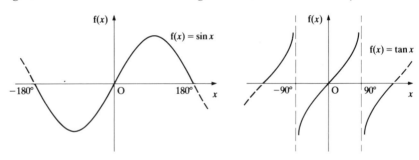

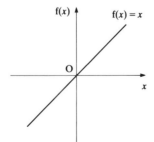
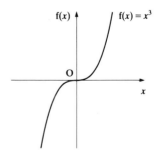

Figure 3.18

In all of these, the left hand half of the graph is obtained from the right hand half by rotating it through $180°$ around the origin.

In such cases,

$$f(-x) = -f(x).$$

EXAMPLE

Show that the function $f(x) = 3x^5 - 2x^3 + x$ is an odd function.

Solution

$$\begin{aligned}
f(-x) &= 3(-x)^5 - 2(-x)^3 + (-x) \\
&= -3x^5 + 2x^3 - x \\
&= -(3x^5 - 2x^3 + x) \\
&= -f(x).
\end{aligned}$$

The function is an odd function.

Any polynomial function $f(x)$ containing only odd powers of x is an odd function.

Not all functions can be classified as even or odd — in fact the majority are neither.

EXAMPLE

For each of the graphs sketched below, say whether the function is odd, even or neither.

(a) 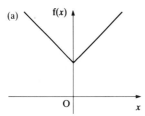 (b) 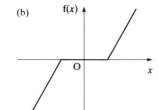 (c)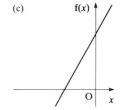

Solution

(a) The graph is symmetrical about the y axis, therefore the function is even.

(b) A rotation of 180° about the origin leaves the graph unchanged, therefore the function is odd.

(c) The graph is changed by a rotation of 180° about the origin, and the y axis is not a line of symmetry, therefore the function is neither odd nor even.

Periodic functions

A *periodic function* is one whose graph has a repeating pattern, just as a periodic sequence is a sequence which repeats itself at regular intervals. You have already met the most common periodic functions – the trigonometrical functions such as $f(x) = \sin x$ (figure 3.19).

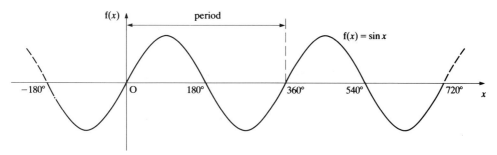

Figure 3.19

A periodic function f(x) is such that there is some value of k for which

$$f(x + k) = f(x) \qquad \text{for all values of } x.$$

The smallest value of k for which this is true is called the *period* of the function.

The functions $f(x) = \sin x$ and $f(x) = \cos x$ both have a period of 360° (or 2π), and $f(x) = \tan x$ has a period of 180° (or π).

EXAMPLE (i)   Sketch the curve of the function $f(x) = 3 \sin (2x - 30°)$.
(ii) State the period of this function.

Solution

(i)

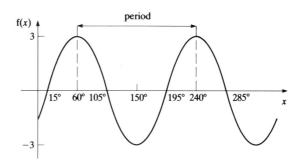

(ii)   Period = 180°

___

You can draw the graph of a periodic function if you know its behaviour over one period.

EXAMPLE The function f(x) is periodic with period 2. Given that

$$f(x) = x^2 \qquad 0 \le x < 1$$
$$f(x) = 2 - x \qquad 1 \le x < 2,$$

sketch the graph of f(x) for $-2 \le x < 4$.

Solution

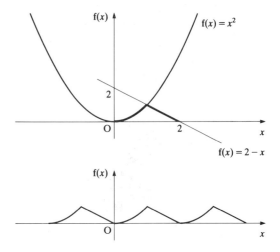

The first diagram shows the parts of the line and the curve which define
f(x). These parts span an interval of length 2 (the period of the function)
and thus form the basic repeating pattern. The second diagram shows this
pattern repeated three times in the interval $-2 \le x < 4$.

Exercise 3E

1. For each of the following curves, say whether the function is odd,
even or neither.

(a)
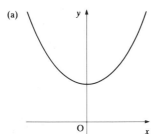

(b)
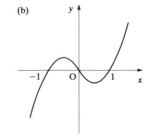

(c)
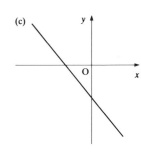

(d)
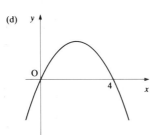

(e)
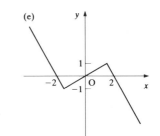

(f)
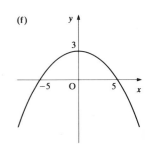

2. For each of the following functions, say whether it is odd, even, periodic, or any combination of these. For any function that is periodic, find its period.

(a) $f(x) = 2 - x^2$ (b) $f(x) = \sin 3x$

(c) $f(x) = x^2 + 2x - 3$ (d) $f(x) = 2x^3 - 3x$

(e) $f(x) = \sin x + \cos x$ (f) $f(x) = \sin x \cos x$

3. (i) Sketch the function $f(x) = \sin 2x$ for $0° \leq x \leq 360°$ and hence state its period.

(ii) Say how the period of this function is related to the period of $\sin x$.

(iii) What are the periods of the following functions?

    (a) $f(x) = \sin 4x$ (b) $f(x) = \sin 3x$ (c) $f(x) = \sin \frac{x}{2}$

4. The function f is even, periodic with period 2, and for $0 \leq x \leq 1$, $f(x) = x$. Sketch the graph of $f(x)$ for $-4 \leq x \leq 4$.

5. A function f has domain the set of real numbers. For $0 \leq x \leq 1$, it is given by the equation

$$f(x) = 1 - x.$$

Given also that f is an even function with period 2, draw its graph over the interval $-3 \leq x \leq 3$.

Write down equations for the function for

(i) $-1 \leq x \leq 0$ (ii) $2 \leq x \leq 3$

[SMP]

6. A function $g(x)$ of period 2 is defined by

$$g(x) = x^2 \qquad \text{for } 0 \leq x \leq \tfrac{1}{2}$$
$$g(x) = \tfrac{1}{4} \qquad \text{for } \tfrac{1}{2} \leq x \leq 1.$$

Given also that $g(x) = g(-x)$ for all x, sketch the graph of $g(x)$ for $-2 \leq x \leq 2$.

7. A light elastic string is stretched between two points A and B which are 3 metres apart on a smooth horizontal surface. A heavy object attached to the midpoint of the string is pulled 50 cm towards A and then released.

During the subsequent motion the string remains taut, and the object oscillates along part of the line AB in such a way that its displacement x cm from the centre of AB at a time t seconds after the motion commences is given by

Exercise 3E continued

$$x = 0.5 \cos 2.5t \qquad \text{where the angle is in radians.}$$

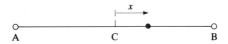

Sketch the graph of x against t for $0 \le t \le 2\pi$.

Hence show that the motion is periodic, and state its period.

Curve sketching

You have already had some experience of curve sketching, and have probably realised that it is of fundamental importance in mathematics. Throughout this course, the curve sketching techniques available to you will be progressively extended. This section reviews the techniques you have met so far.

When sketching a curve you need to mark points in approximately the right positions and join them up in the right general shape. You should also indicate the co-ordinates of any important points, such as points of intersection with the axes and turning points. Your sketch should show up any symmetry which the curve possesses, any asymptotes, and should indicate the behaviour of the curve for large values of x or y.

A graphics calculator or suitable computer software is often useful, but you must be careful that important features of a graph are not missed. This happens most often when either a turning point is off the screen with the range which is being used, or two or more turning points are so close together that they cannot be distinguished. The following investigation shows how this can happen, and suggests some questions which you should ask yourself before you accept the graph that is displayed.

Activity

Use a graphics calculator or computer program with the range set at x min: -2, x max: 2, y min: -10, y max: 20.

1. Sketch the graph of $f(x) = x^3 - 15x^2 + 27x + 1$
  (i) How many turning points are there in the display?
  (ii) The function is a cubic function. How many turning points might you expect?
  (iii) The function has a positive x^3 term. What would you expect for the general shape of the curve?
  (iv) Use your answers to (ii) and (iii) to alter the range so that you obtain a true picture of the function.

2. Sketch the graph of $f(x) = 10x^4 - x^2 + 1$

    (i)   How many turning points seem to be in the display?

    (ii) The function is a quartic (fourth degree) function. How many turning points might you expect?

    (iii) The function has a positive x^4 term. What would you expect for the general shape of the curve?

    (iv) Use your answers to (ii) and (iii) to alter the range (or zoom in) so that you obtain a more detailed picture of the function.

3. Sketch the graph of $f(x) = \dfrac{8x + 3}{x - 5}$.

    (i)   The display obviously shows only part of the curve. The function has the term $x - 5$ in the denominator: which value of x must therefore be excluded from the domain?

    (ii) Alter the range setting so that this value is visible. You will find that altering it just to include this value tells you very little more about the curve – you need to make quite considerable alterations to get a good idea of the correct graph. Possible settings are x min: -10, x max: 20, y min: -20, y max: 40.

Earlier in this chapter you sketched curves by relating their equations to those of standard curves. This process gives you the shape of the curve, but further calculation may be needed, for example to find where the curve crosses the axes. In cases where it is not possible to use transformations you need an alternative strategy. The following checklist may be helpful.

1.   **Find where the curve cuts the axes.**

    The graph of $y = f(x)$ cuts the y axis when $x = 0$.

    It cuts the x axis when $y = 0$, i.e. when x is such that $f(x) = 0$. Only calculate these points if this is reasonably simple.

2.   **Check for any obvious symmetry.**

    Is the function odd, even or periodic?

3.   **Find any vertical asymptotes.**

    e.g. $x = 0$ is an asymptote for $f(x) = \frac{1}{x}$

             $x = 2$ is an asymptote for $f(x) = 3 + \dfrac{5}{x - 2}$

             $\theta = 90°$ is an asymptote for $f(\theta) = \tan \theta$

    Asymptotes other than the co-ordinate axes are always shown on a sketch using a dotted line.

4.   **Examine the behaviour as $x \to \pm \infty$**

    The function may approach a constant value as x tends to $\pm\infty$, and in that case the curve will have a horizontal asymptote. Figure 3.20(a) shows such a function, with a horizontal asymptote at $y = 2$. Alternatively, one of the terms in the function may become dominant, so that it determines the shape of the curve for large values of x (positive or negative). Figure 3.20(b) shows a curve of this type: the term x dominates $\frac{1}{x}$ for large x.

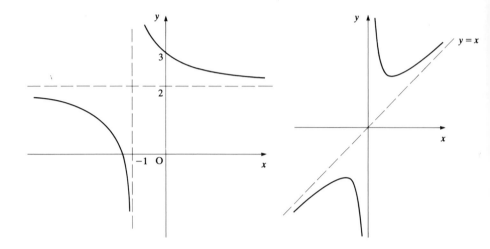

(a) $y = \dfrac{2x + 3}{x + 1}$

(b) $y = x + \dfrac{1}{x}$

Figure 3.20

5. **Look for any stationary points.**

   To do this, use the calculus methods that you met in Pure Mathematics 1.

Exercise 3F

1. The diagram shows the graph of

$$f(x) = x^4 - 8x^2 + 10.$$

State the co-ordinates of the points A and B, giving reasons for your answers.

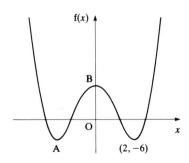

2. The diagram shows the graph of

$$f(x) = \frac{1}{(x+2)(x-2)}$$

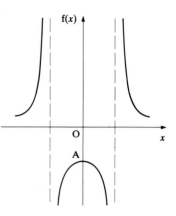

State

    (i) the co-ordinates of point A.

    (ii) the equations of the asymptotes.

3. The function $f(x)$ is defined by $f(x) = \dfrac{x+2}{x-1}$ $x \in \mathbb{R},\, x \neq 1$.

Write this in the form $f(x) = a + \dfrac{b}{x-1}$, where a and b are constants to be determined. Find

  (i) the equations of the asymptotes

  (ii) the co-ordinates of the points where the curve crosses the axes.

Sketch the curve.

4. The function $f(x)$ is defined by $f(x) = x^2 + \dfrac{8}{x^2}$ $x \in \mathbb{R},\, x \neq 0$.

  (i) State the equation of the vertical asymptote.

  (ii) Identify any symmetries which the curve possesses.

  (iii) What can you say about the sign of $f(x)$?

  (iv) Describe the behaviour of the curve $y = f(x)$ for large positive or negative values of x.

  (v) Sketch the curve $y = x^2$.

  (vi) On the same graph, sketch the curve $y = f(x)$, paying particular attention to any relationship between the two curves.

5. (i) Write the equation $y = \dfrac{2x-3}{x+1}$ in the form $y = \dfrac{a}{x+1} + b$ where a and b are constants to be determined.

  (ii) Hence sketch the curve $y = \dfrac{2x-3}{x+1}$.

Investigation

Investigate the relationship between the graphs of $y = f(x)$ and $y = \dfrac{1}{f(x)}$ for different functions $f(x)$.

KEY POINTS

Mappings and functions

- A mapping is any rule connecting input values (objects) and output values (images). It can be many-to-one, one-to-many, one-to-one or many-to-many.

- A many-to-one or one-to-one mapping is called a function. It is a mapping for which each input value gives exactly one output value.

- The domain of a mapping or function is the set of possible input values (values of x).

- The co-domain of a mapping or function is the set of possible output values (values of y).

- The range of a mapping or function is the set of output values which are actually achieved.

Transformations of the graphs of the function $y = f(x)$

Function	Transformation
$f(x - t) + s$	Translation $\begin{pmatrix} t \\ s \end{pmatrix}$
$a\,f(x)$	One way stretch, parallel y axis, scale factor a
$f(ax)$	One way stretch, parallel x axis scale factor $\frac{1}{a}$
$-f(x)$	Reflection in x axis
$f(-x)$	Reflection in y axis

Composite functions

- A composite function is obtained when one function (say g) is applied after another (say f). Notation is g[f(x)] or gf(x).

(cont. . .)

Inverse functions

- For any one-to-one function $f(x)$, there is an inverse function $f^{-1}(x)$.

- The curves of a function and its inverse are reflections of each other in the line $y = x$.

Special functions

- An even function: $f(x) = f(-x)$: the y axis is a line of symmetry

- An odd function: $f(x) = -f(-x)$: rotational symmetry about the origin.

- A periodic function: $f(x + k) = f(x)$: a repeating pattern of length k.

4

Calculus techniques

To find the simple in the complex, the finite in the infinite – that is not a bad description of mathematics.

Jacob Schwartz

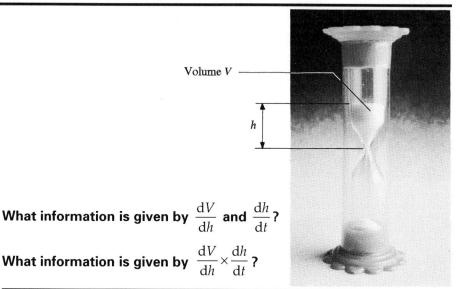

Volume V

h

What information is given by $\dfrac{dV}{dh}$ **and** $\dfrac{dh}{dt}$ **?**

What information is given by $\dfrac{dV}{dh} \times \dfrac{dh}{dt}$ **?**

You have already seen how differentiating a function to find any stationary points helps you to sketch the curve of the function. However, the number of functions that you can differentiate at the moment is rather limited, since the only rule you have used is

$$y = kx^n \quad \Rightarrow \quad \frac{dy}{dx} = knx^{n-1} \qquad \text{where } k \text{ is a constant and } n \text{ a positive integer.}$$

The work in this chapter enables you to differentiate many more functions.

Look at the function $f(x) = \frac{1}{x}$ (i.e. $y = x^{-1}$) where $x \neq 0$. Since n is negative in this case, we must start from first principles.

$$\frac{dy}{dx} = \lim_{\delta x \to 0} \frac{f(x + \delta x) - f(x)}{\delta x}$$

$$= \lim_{\delta x \to 0} \frac{\dfrac{1}{x + \delta x} - \dfrac{1}{x}}{\delta x} \qquad \text{For } y = \frac{1}{x}$$

$$= \lim_{\delta x \to 0} \frac{x - (x + \delta x)}{x(x + \delta x)\delta x}$$

$$= \lim_{\delta x \to 0} \frac{-\delta x}{x(x + \delta x)\delta x}$$

$x + \delta x \to x$

$$= \lim_{\delta x \to 0} \frac{-1}{x(x + \delta x)}$$

In the limit as $\delta x \to 0$

$$= -\frac{1}{x^2}$$

So, $y = x^{-1} \quad \Rightarrow \quad \dfrac{dy}{dx} = x^{-2}, \quad x \neq 0.$

Notice that this is the result you would have obtained using the existing rule if you put $n = -1$. This suggests that the existing rule may be extended to other values of n. This is indeed true and the rule applies for all real values of n, including negative numbers and fractions.

EXAMPLE Use the extended rule to differentiate the following functions.

(a) $y = \dfrac{1}{x^2}$ (b) $y = \sqrt{x}$

Solution

(a) $y = \dfrac{1}{x^2} = x^{-2}$

$\dfrac{dy}{dx} = -2 \times x^{-2-1}$

$= -2x^{-3}$

$= \dfrac{-2}{x^3}$

(b) $y = \sqrt{x} = x^{1/2}$

$\dfrac{dy}{du} = \dfrac{1}{2} \times x^{1/2 - 1}$

$= \dfrac{1}{2} x^{-1/2}$

$= \dfrac{1}{2\sqrt{x}}$

EXAMPLE Given that $y = x \sqrt{x}$ find $\dfrac{dy}{dx}$, and evaluate $\dfrac{dy}{dx}$ at the point $(9, 27)$.

Solution

$$y = x \sqrt{x} = x^{3/2}$$

$$\frac{dy}{dx} = \frac{3}{2} \times x^{3/2 - 1}$$

$$= \frac{3}{2} x^{1/2}$$

When $x = 9$, $\dfrac{dy}{dx} = \dfrac{3}{2} \times 9^{\frac{1}{2}}$

$$= \dfrac{3}{2} \times 3$$

$$= 4\tfrac{1}{2}$$

Exercise 4A

In some of these questions you will find it useful to use a graphics calculator to check your answers.

1. Differentiate the following functions using the rules

$$y = kx^n \quad \Rightarrow \quad \dfrac{dy}{dx} = knx^{n-1} \quad \text{(for } k \text{ any constant, } n \text{ any real number)}$$

and $y = \text{f}(x) + \text{g}(x) \quad \Rightarrow \quad \dfrac{dy}{dx} = \text{f}'(x) + \text{g}'(x).$

(a) $y = \dfrac{1}{x^4}$

(b) $y = 4x^{-5}$

(c) $y = 7x^{-6}$

(d) $y = \dfrac{3}{x^2}$

(e) $y = \dfrac{3}{x^5}$

(f) $y = \dfrac{2}{x} + x^3$

(g) $y = \dfrac{5}{x^3} - \dfrac{2}{x} + 1$

(h) $y = x^{\frac{1}{4}}$

(i) $y = 6x^{\frac{1}{3}}$

(j) $y = \sqrt[5]{x}$

(k) $y = \sqrt{x} + \dfrac{1}{x^3} + 2x$

(l) $y = \left(\sqrt[3]{x}\right)^4$

2. For each part of this question,

(i) find $\dfrac{dy}{dx}$, and (ii) find the gradient of the curve at the given point.

(a) $y = x^{-2}$; $(0.25, 16)$

(b) $y = x^{-1} + x^{-4}$; $(-1, 0)$

(c) $y = 4x^{-3} + 2x^{-5}$; $(1, 6)$

(d) $y = 3x^4 - 4 - 8x^{-3}$; $(2, 43)$

3. (i) Sketch the curve $y = \dfrac{1}{x} + 2$.

(ii) Write down the co-ordinates of the point where the curve crosses the x axis.

(iii) Differentiate $y = \dfrac{1}{x} + 2$.

(iv) Find the gradient of the curve at the point where it crosses the x axis.

4. The graph of $y = \dfrac{4}{x^2} + x$ is shown below.

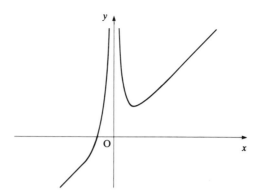

(i)   Differentiate $y = \dfrac{4}{x^2} + x$.

(ii) Show that the point $(-2, -1)$ lies on the curve.

(iii) Find the gradient of the curve at $(-2, -1)$.

(iv) Show that the point $(2, 3)$ lies on the curve.

(v) Find the gradient of the curve at $(2, 3)$.

(vi) Relate your answer to part (v) to the shape of the curve.

5. (i) Plot on the same axes the graphs whose equations are

$$y = \frac{1}{x^2} + 1 \quad \text{and} \quad y = -16x + 13 \quad \text{for} \quad -3 \leqslant x \leqslant 3$$

(ii) Show that the point $(0.5, 5)$ lies on both graphs.

(iii) Differentiate $y = \dfrac{1}{x^2} + 1$ and find its gradient at $(0.5, 5)$.

(iv) What can you deduce about the two graphs?

6. (i)   Sketch the curve $y = \sqrt{x} - 1$.

(ii) Differentiate $y = \sqrt{x} - 1$.

(iii) Find the co-ordinates of the point on the curve $y = \sqrt{x} - 1$ at which the tangent is parallel to the line $y = 2x - 1$.

(iv) Is the line $y = 2x - 1$ a tangent to the curve $y = \sqrt{x} - 1$? Give reasons for your answer.

Exercise 4A continued

7. The graph of $y = 3x - \dfrac{1}{x^2}$ is shown below. The point marked P is $(1, 2)$.

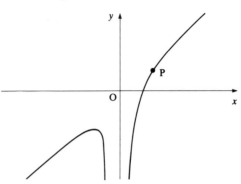

(i) Find the gradient function $\dfrac{dy}{dx}$.

(ii) Use your answer from part (i) to find the gradient of the curve at P.

(iii) Use your answer from part (ii), and the fact that the gradient of the curve at P is the same as that of the tangent at P, to find the equation of the tangent at P in the form $y = mx + c$.

8. The graph of $y = x^2 + \dfrac{1}{x}$ is shown below. The point marked Q is $(1, 2)$.

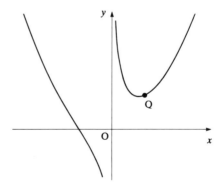

(i) Find the gradient function $\dfrac{dy}{dx}$.

(ii) Find the gradient of the tangent at Q.

(iii) Show that the equation of the normal to the curve at Q can be written as $x + y = 3$.

(iv) At what other points does the normal cut the curve?

Higher derivatives

Figure 4.1 shows a sketch of a function $y = f(x)$, and beneath it a sketch of the corresponding gradient function $\dfrac{dy}{dx} = f'(x)$.

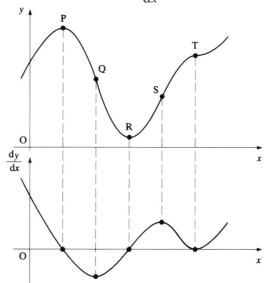

Figure 4.1

Activity

Sketch the graph of the gradient of $\dfrac{dy}{dx}$ against x for the function illustrated above. Do this by tracing the two graphs shown, and extending the dashed lines downwards onto a third set of axes.

You can see that P is a maximum point, R is a minimum point and T is a stationary point of inflection, and that Q and S are non-stationary points of inflection. What can you say about the gradient of $\dfrac{dy}{dx}$ at these points: is it positive, negative or zero?

The gradient of any point on the curve of $\dfrac{dy}{dx}$ is given by $\dfrac{d}{dx}\left(\dfrac{dy}{dx}\right)$. This is written as $\dfrac{d^2y}{dx^2}$ or $f''(x)$, and is called the second derivative. It is found by differentiating the function a second time.

NOTE

The second derivative, $\dfrac{d^2y}{dx^2}$, is not the same as $\left(\dfrac{dy}{dx}\right)^2$

EXAMPLE

Given that $y = x^5 + 2x$, find $\dfrac{d^2y}{dx^2}$

Solution

$$\frac{dy}{dx} = 5x^4 + 2$$

$$\frac{d^2y}{dx^2} = 20x^3$$

Using the second derivative

You can use the second derivative to identify the nature of a stationary point, instead of looking at the sign of $\frac{dy}{dx}$ just either side of it.

Notice that at P, $\frac{dy}{dx} = 0$ and $\frac{d^2y}{dx^2} < 0$. This tells you that the gradient, $\frac{dy}{dx}$, is zero and decreasing. It must be going from positive to negative, so P is a maximum point (figure 4.2).

At R, $\frac{dy}{dx} = 0$ and $\frac{d^2y}{dx^2} > 0$. This tells you that the gradient, $\frac{dy}{dx}$ is zero and increasing. It must be going from negative to positive, so R is a minimum point (figure 4.3)

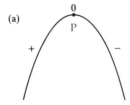

Figure 4.2   **Figure 4.3**

The next examples illustrates the use of the second derivative to identify the nature of stationary points.

EXAMPLE   Given that $y = 2x^3 + 3x^2 - 12x$,

(i)   find $\frac{dy}{dx}$, and find the values of x for which $\frac{dy}{dx} = 0$;

(ii) find the value of $\frac{d^2y}{dx^2}$, at each stationary point and hence determine its nature;

(iii) find the y values of each of the stationary points;

(iv) sketch the curve given by $y = 2x^3 + 3x^2 - 12x$.

Solution

(i) $\dfrac{dy}{dx} = 6x^2 + 6x - 12$

$\qquad = 6(x^2 + x - 2)$

$\qquad = 6(x + 2)(x - 1)$

$\dfrac{dy}{dx} = 0$ when $x = -2$ or 1

(ii) $\dfrac{d^2y}{dx^2} = 12x + 6$

$\qquad = 6(2x + 1).$

When $x = -2$, $\dfrac{d^2y}{dx^2} = 6(2 \times (-2) + 1) = -18.$

$\dfrac{d^2y}{dx^2} < 0 \quad \Rightarrow \quad$ a maximum

When $x = 1$, $\dfrac{d^2y}{dx^2} = 6(2 \times 1 + 1) = 18$

$\dfrac{d^2y}{dx^2} > 0 \quad \Rightarrow \quad$ a minimum

(iii) When $x = -2$, $y = 2(-2)^3 + 3(-2)^2 - 12(-2)$

$\qquad\qquad\qquad = 20,$

so $(-2, 20)$ is a maximum point.

When $x = 1$, $y = 2 + 3 - 12$

$\qquad\qquad\quad = -7,$

so $(1, -7)$ is a minimum point.

(iv)

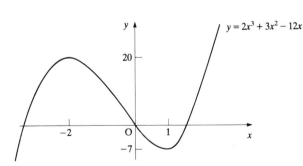

$y = 2x^3 + 3x^2 - 12x$

NOTE

On occasions when it is difficult or laborious to find $\dfrac{d^2y}{dx^2}$, remember that you can always classify the nature of a stationary point by looking at the sign of $\dfrac{dy}{dx}$ for points just either side of it.

Points of inflection

For Discussion

The map shows the motor-racing circuit at Brands Hatch. At what points of the circuit are there points of inflection, and how can the drivers know when they are at one?

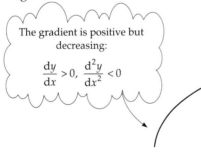

When a curve is drawn on Cartesian axes, the points of inflection may be divided into two categories: non-stationary and stationary.

At a non-stationary point of inflection, $\dfrac{d^2y}{dx^2} = 0$ but $\dfrac{dy}{dx} \neq 0$ as shown in figure 4.4.

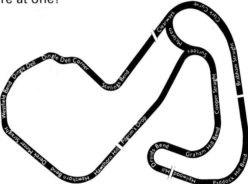

The gradient is positive but decreasing:
$$\frac{dy}{dx} > 0, \ \frac{d^2y}{dx^2} < 0$$

The gradient is positive and increasing:
$$\frac{dy}{dx} > 0, \ \frac{d^2y}{dx^2} > 0$$

At the point of inflection the gradient is positive but neither increasing nor decreasing:
$$\frac{dy}{dx} > 0, \ \frac{d^2y}{dx^2} = 0$$

Figure 4.4

Activity

Draw a diagram like the one above to illustrate a non-stationary point of inflection on a decreasing curve, commenting on the gradient and how it is changing before, at and after the point of inflection.

At a stationary point of inflection, both $\dfrac{dy}{dx}$ and $\dfrac{d^2y}{dx^2}$ equal zero, as in the case of $y = x^3$ at the origin (figure 4.5).

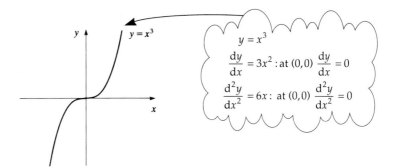

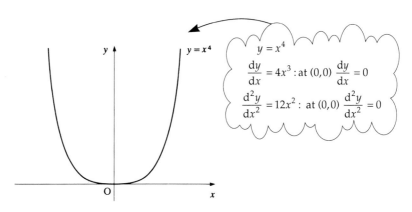

Figure 4.5

While it is true for all stationary points of inflection that $\dfrac{dy}{dx} = 0$ and $\dfrac{d^2y}{dx^2} = 0$, it is also true for some turning points. For example, the curve $y = x^4$ has a minimum point at the origin (figure 4.6).

Figure 4.6

Consequently, if both $\dfrac{dy}{dx}$ and $\dfrac{d^2y}{dx^2}$ are zero at a point, you still need to check the values of $\dfrac{dy}{dx}$ either side of the point in order to determine its nature.

EXAMPLE   (i) Find and classify any stationary points on the curve $y = x^4 + x^3$.

(ii) Sketch the curve $y = x^4 + x^3$.

Solution

(i) Finding the stationary points $\dfrac{dy}{dx} = 4x^3 + 3x^2$

$$= x^2(4x + 3)$$

For a stationary point, $\dfrac{dy}{dx} = 0 \quad \Rightarrow x = 0 \text{ or } -\dfrac{3}{4}$.

$$x = 0 \Rightarrow y = 0; \quad x = -\tfrac{3}{4} \Rightarrow y = -\tfrac{27}{256}$$

The stationary points are at $(0,0)$ and $\left(-\tfrac{3}{4}, \ -\tfrac{27}{256}\right)$

Classifying the stationary points $\dfrac{d^2y}{dx^2} = 12x^2 + 6x$

When $x = -\dfrac{3}{4}$, $\dfrac{d^2y}{dx^2} > 0$, so there is a minimum point at $\left(-\tfrac{3}{4}, \ -\tfrac{27}{256}\right)$

When $x = 0$, $\dfrac{d^2y}{dx^2} = 0$, so you need to investigate further.

x is just less than $0 \Rightarrow \dfrac{dy}{dx} > 0.$

x is just greater than $0 \Rightarrow \dfrac{dy}{dx} > 0.$

Conclusion: there is a point of inflection at $(0,0)$

(ii)

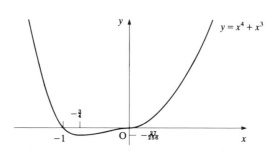

Exercise 4B

1. For each of the following functions, find $\dfrac{dy}{dx}$ and $\dfrac{d^2y}{dx^2}$.

(a) $y = x^3$

(b) $y = x^5$

(c) $y = 4x^2$

(d) $y = x^{-2}$

(e) $y = x^{\frac{3}{2}}$

(f) $y = x^4 - \dfrac{2}{x^3}$

2. Find any stationary points on the curves of the following functions and identify their nature. You do not need to sketch the curves but you should check your answers, if possible, using a graphics calculator.

(a) $y = x^2 + 2x + 4$

(b) $y = 6x - x^2$

(c) $y = x^3 - 3x$

(d) $y = 4x^5 - 5x^4$

(e) $y = x + \dfrac{1}{x}$

(f) $y = x^3 + \dfrac{12}{x}$

(g) $y = 6x - x^{\frac{3}{2}}$

(h) $y = x^4 + x^3 - 2x^2 - 3x + 1$

3. Given that $y = x^4 - 8x^2$

(i) find $\dfrac{dy}{dx}$;

(ii) find $\dfrac{d^2y}{dx^2}$;

(iii) Find any stationary points and identify their nature.

(iv) Hence sketch the curve.

4. Charlie wants to add an extension with a floor area of 18 m^2 to the back of his house. He wants to use the minimum possible number of bricks, so he wants to know the smallest perimeter he can use. The dimensions, in metres, are x and y as shown.

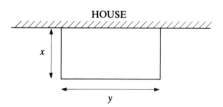

(i) Write down an expression for the area in terms of x and y.

(ii) Write down an expression, in terms of x and y, for the perimeter, P, of the outside walls.

(iii) Show that

$$P = 2x + \dfrac{18}{x}.$$

Exercise 4B continued

(iv) Find $\dfrac{\mathrm{d}P}{\mathrm{d}x}$ and $\dfrac{\mathrm{d}^2P}{\mathrm{d}x^2}$.

(v) Find the dimensions of the extension that give a minimum perimeter, and confirm that it is a minimum.

5. A fish tank with a square base and no top is to be made from a thin sheet of toughened glass. The dimensions are as shown.

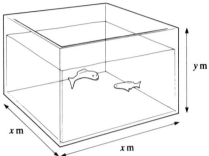

(i) Write down an expression for the volume V in terms of x and y.
(ii) Write down an expression for the total surface area A in terms of x and y.

The tank needs a capacity of 0.5 m^3 and the manufacturer wishes to use the minimum possible amount of glass.

(iii) Deduce an expression for A in terms of x only.
(iv) Find $\dfrac{\mathrm{d}A}{\mathrm{d}x}$ and $\dfrac{\mathrm{d}^2A}{\mathrm{d}x^2}$.

(v) Find the values of x and y that use the smallest amount of glass and confirm that these give the minimum value.

6. A closed rectangular box is made of thin card, and its length is 3 times its width. The height is h cm and the width is x cm.

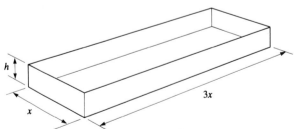

(i) The volume of the box is 972 cm^3. Use this to write down an expression for h in terms of x.
(ii) Show that the surface area, A, can be written as $A = 6x^2 + \dfrac{2592}{x}$.

    (iii) Find $\dfrac{\mathrm{d}A}{\mathrm{d}x}$ and use it to find a stationary point. Find $\dfrac{\mathrm{d}^2A}{\mathrm{d}x^2}$ and use it to verify that the stationary point gives the minimum value of A.

    (iv) Hence find the minimum surface area and the corresponding dimensions of the box.

7. A cylindrical can with a lid is to be made from a thin sheet of metal. Its height is to be h cm and its radius r cm . The surface area is to be 250π cm^3.

    (i) Find h in terms of r.

    (ii) Write down an expression for the volume, V, of the can in terms of r.

    (iii) Find $\dfrac{\mathrm{d}V}{\mathrm{d}r}$ and $\dfrac{\mathrm{d}^2V}{\mathrm{d}r^2}$

    (iv) Use your answers from (iii) to show that the can's maximum possible volume is 1690 cm^3 (to 3 significant figures), and find the corresponding dimensions of the can.

8. A garden is planned with a lawn area of 24 m^2 and a path around the edge. The dimensions of the lawn and path are as shown in the diagram.

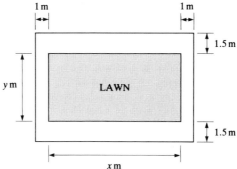

    (i) Write down an expression, for y in terms of x.

    (ii) Find an expression for the overall area of the garden, A, in terms of x.

    (iii) Find the smallest possible overall area for the garden.

9. An open cylindrical rubbish bin with volume $\frac{\pi}{8}$ m^3 is to be made from a large sheet of metal. To keep costs as low as possible, the bin should have the minimum possible surface area, A m^2.

    (i) Using the fact that the volume is $\frac{\pi}{8}$ m^3, write down an expression for the height, h, in terms of the radius, r (both in metres).

    (ii) Use this expression to show that A can be written as:

$$A = \pi r^2 + \frac{\pi}{4r}$$

Exercise 4B continued

    (iii) Find $\dfrac{\mathrm{d}A}{\mathrm{d}r}$ and $\dfrac{\mathrm{d}^2A}{\mathrm{d}r^2}$, and hence find the dimensions which make the area a minimum.

    (iv) Find the corresponding minimum value.

10. A ship is to make a voyage of 100 km at a constant speed of v kmh^{-1}.

The running cost of the ship is $£\left(0.8v^2 + \dfrac{2000}{v}\right)$ per hour, and this is to be kept to minimum.

    (i) Write down the time taken to go 100 km at v kmh^{-1}.

    (ii) Hence write down the total cost, $£\,C$, of travelling 100 km at v kmh^{-1}.

    (iii) By considering $\dfrac{\mathrm{d}C}{\mathrm{d}v}$ find the speed which keeps the cost of the journey to a minimum. (Give your answer to the nearest kmh^{-1}.)

    (iv) Find the minimum cost of the voyage.

11. A cylindrical tank, open at the top, has height h m and radius r m. It has a capacity of 1 m^3.

    (i) Show that $h = \dfrac{1}{\pi r^2}$.

    (ii) Its total internal surface area is S m^2.

    (iii) Show that $S = \dfrac{2}{r} + \pi r^2$

Determine the value of r which makes the surface area S as small as possible.

[MEI]

12. The diagram shows a right-angled triangle with an area of 8 cm^2.

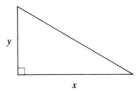

    (i) Write down an expression for y in terms of x.

    (ii) Write down an expression for the sum, S, of the squares of these two numbers in terms of x.

    (iii) Find the least value of the sum of their squares by considering $\dfrac{\mathrm{d}S}{\mathrm{d}x}$ and $\dfrac{\mathrm{d}^2S}{\mathrm{d}x^2}$.

    (iv) Hence write down the shortest possible length for the hypotenuse.

The chain rule

How would you differentiate an expression like

$$y = \sqrt{(x^2 + 1)} \,?$$

Your first thought may be to write it as $y=(x^2+1)^{\frac{1}{2}}$ and then get rid of the brackets, but that is not possible in this case because the power $\frac{1}{2}$ is not a positive integer. Instead you need to think of the expression as a composite function, a 'function of a function'.

You have already met composite functions in chapter 3, using the notation $g[f(x)]$ or $gf(x)$.

In this chapter we call the first function to be applied $u(x)$, or just u, rather than $f(x)$.

In this case, $u=x^2+1$
and $y= \sqrt{u}=u^{\frac{1}{2}}$.

This is now in a form which you can differentiate using the *chain rule*.

Differentiating a composite function

To find $\dfrac{dy}{dx}$ for a function of a function, you consider the effect of a small change in x on the two variables, y and u, as follows. A small change δx in x leads to a small change δu in u and a corresponding small change δy in y, and by simple algebra,

$$\frac{\delta y}{\delta x} = \frac{\delta y}{\delta u} \times \frac{\delta u}{\delta x}.$$

In the limit, as $\delta x \to 0$,

$$\frac{\delta y}{\delta x} \to \frac{dy}{dx}, \quad \frac{\delta y}{\delta u} \to \frac{dy}{du} \text{ and } \frac{\delta u}{\delta x} \to \frac{du}{dx}$$

and so the relationship above becomes

$$\frac{dy}{dx} = \frac{dy}{du} \times \frac{du}{dx}$$

This is known as the chain rule.

EXAMPLE   Differentiate $y = (x^2 + 1)^{\frac{1}{2}}$.

Solution

As you saw earlier, you can break down this expression as follows:

$$y = u^{\frac{1}{2}}, \qquad u = x^2 + 1.$$

Differentiating these gives

$$\frac{dy}{du} = \frac{1}{2}u^{-1/2} = \frac{1}{2\sqrt{(x^2 + 1)}} \qquad \text{and} \qquad \frac{du}{dx} = 2x.$$

By the chain rule

$$\frac{dy}{dx} = \frac{dy}{du} \times \frac{du}{dx}$$

$$= \frac{1}{2\sqrt{(x^2 + 1)}} \times 2x$$

$$= \frac{x}{\sqrt{(x^2 + 1)}}$$

Notice that the answer must be given in terms of the same variables as the question, in this case x and y. The variable u was our invention and so should not appear in the answer.

You can see that effectively you have made a substitution, in this case $u = x^2 + 1$. This transformed the problem into one that could easily be solved. Notice that the substitution gave you two functions that you could differentiate. Some substitutions would not have worked. For example, the substitution $u = x^2$, would give you

$$y = (u + 1)^{\frac{1}{2}} \text{ and } u = x^2$$

You would still not be able to differentiate y, so you would have gained nothing.

EXAMPLE Use the chain rule to find $\dfrac{dy}{dx}$ when $y = (x^2 - 2)^6$

Solution

Let $u = x^2 - 2$, then $y = u^6$.

$$\frac{du}{dx} = 2x \qquad \text{and} \qquad \frac{dy}{du} = 6u^5$$

$$= 6(x^2 - 2)^5$$

$$\frac{dy}{dx} = \frac{dy}{du} \times \frac{du}{dx}$$

$$= 6(x^2 - 2)^5 \times 2x$$

$$= 12x(x^2 - 2)^5.$$

NOTE *With practice you may find that you can do some stages of questions like this in your head, and just write down the answer. If you have any doubt, however, you should write down the full method.*

Differentiation with respect to different variables

The chain rule makes it possible to differentiate with respect to a variable which does not feature in the original expression. For example, the volume V of a sphere of radius r is given by $V = \frac{4}{3}\pi r^3$. Differentiating this with respect to r gives the rate of change of volume with radius, $\dfrac{dV}{dr} = 4\pi r^2$.

However you might be more interested in finding $\dfrac{dV}{dt}$, the rate of change of volume with time, t.

To find this, you would use the chain rule:

$$\frac{dV}{dt} = \frac{dV}{dr} \times \frac{dr}{dt}$$

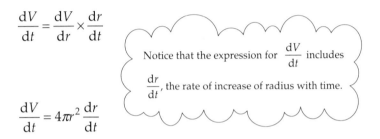

Notice that the expression for $\dfrac{dV}{dt}$ includes $\dfrac{dr}{dt}$, the rate of increase of radius with time.

$$\frac{dV}{dt} = 4\pi r^2 \frac{dr}{dt}$$

You have now differentiated V with respect to t.

The use of the chain rule in this way widens the scope of differentiation and this means that you have to be careful how you describe the process. "Differentiate $y = x^2$" could mean differentiation with respect to x, or t, or any other variable. In this book, and others in this series, we have adopted the convention that, unless we state otherwise, differentiation is with respect to the variable on the right hand side of the expression. So when we write "Differentiate $y = x^2$" or simply "Differentiate x^2", it is to be understood that the differentiation is with respect to x.

Exercise 4C

In some of these questions you are asked to find the stationary points of a curve and then to use them as a guide for sketching it. You will find it helpful to use a graphics calculator to check your answers in these cases.

1. Use the chain rule to differentiate the following functions.

   (a) $y = (x + 2)^3$ (b) $y = (2x + 3)^4$ (c) $y = (x^2 - 5)^3$

   (d) $y = (x^3 + 4)^5$ (e) $y = (3x + 2)^{-1}$ (f) $y = \dfrac{1}{(x^2 - 3)^3}$

   (g) $y = (x^2 - 1)^{\frac{3}{2}}$ (h) $y = \left(\dfrac{1}{x} + x\right)^3$ (i) $y = (\sqrt{x} - 1)^4$

Exercise 4C continued

2. Given that $y = (3x - 5)^3$

(i) find $\dfrac{dy}{dx}$;

(ii) find the equation of the tangent to the curve at $(2,1)$;

(iii) show that the equation of the normal to the curve at $(1,-8)$ can be written in the form

$$36y + x + 287 = 0$$

3. Given that $y = (2x - 1)^4$,

(i) find $\dfrac{dy}{dx}$;

(ii) find the co-ordinates of any stationary points and determine their nature;

(iii) sketch the curve.

4. Given that $y = (x^2 - 4)^3$,

(i) find $\dfrac{dy}{dx}$;

(ii) find the co-ordinates of any stationary points and determine their nature;

(iii) sketch the curve.

5. Given that $y = (x^2 - x - 2)^4$,

(i) find $\dfrac{dy}{dx}$;

(ii) find the co-ordinates of any stationary points and determine their nature;

(iii) sketch the curve.

6. The graph of $y = (x^3 - x^2 + 2)^3$, is shown in the diagram.

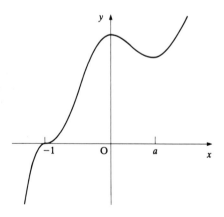

Exercise 4C continued

   (i) Find the gradient function $\dfrac{dy}{dx}$.

   (ii) Verify, showing your working clearly, that when $x = -1$ the curve has a point of inflection and when $x = 0$ the curve has a maximum.

   (iii) The curve has a minimum when $x = a$. Find a and verify that this corresponds to a minimum.

   (iv) Find the gradient at $(1, 8)$ and the equation of the tangent to the curve at this point.

7. Some students on an expedition reach the corner of a very muddy field. They need to reach the opposite corner as quickly as possible as they are behind schedule. They estimate that they could walk along the edge of the field at 5 kmh^{-1} and across the field at 3 kmh^{-1}. They know from their map that the field is a square of side 0.5 km.

How far should they walk along the edge of the field before cutting across?

The product rule

Figure 4.7 shows a sketch of the curve of $y = 20x\,(x - 1)^6$.

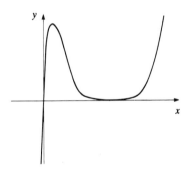

Figure 4.7

If you wanted to find the gradient function, $\dfrac{dy}{dx}$, for the curve, you could expand the right hand side then differentiate it term by term – a long and cumbersome process!

There are other functions like this, made up of the product of two or more simpler functions, which are not just time-consuming to expand – they are *impossible* to expand. One such function is

$$y = (x - 1)^{\frac{1}{2}}\,(x + 1)^6$$

Clearly you need a technique for differentiating functions which are products of simpler ones, and a suitable notation with which to express it.

The most commonly used notation involves writing

$$y = uv$$

where the variables u and v are both functions of x. Using this notation, $\dfrac{dy}{dx}$ is given by

$$\frac{dy}{dx} = u\frac{dv}{dx} + v\frac{du}{dx}.$$

This is called the *product rule* and it is derived from first principles in the next section.

The product rule from first principles

A small increase δx in x leads to corresponding small increases δu, δv and δy, in u, v and y. And so

$$y + \delta y = (u + \delta u)(v + \delta v)$$

$$= uv + v\delta u + u\delta v + \delta u\delta v$$

Since $y = uv$, the increase in y is given by

$$\delta y = v\delta u + u\delta v + \delta u\delta v.$$

Dividing both sides by δx,
$$\frac{\delta y}{\delta x} = v\frac{\delta u}{\delta x} + u\frac{\delta v}{\delta x} + \delta u\frac{\delta u}{\delta x}$$

In the limit, as $\delta x \to 0$, so do δu, δv and δy, and

$$\frac{\delta u}{\delta x} \to \frac{du}{dx}, \qquad \frac{\delta v}{\delta x} \to \frac{dv}{dx} \qquad \text{and} \qquad \frac{\delta y}{\delta x} \to \frac{dy}{dx}$$

The expression becomes
$$\frac{dy}{dx} = v\frac{du}{dx} + u\frac{dv}{dx}$$

Notice that since $\delta u \to 0$ the last term on the right hand side has disappeared.

EXAMPLE

Given that $y = (3x + 2)(x^2 - 5)$, find $\dfrac{dy}{dx}$ using the product rule.

Solution

$$y = (2x + 3)(x^2 - 5)$$

Let $u = 2x + 3$ and $v = x^2 - 5$

Then $\dfrac{du}{dx} = 2$ and $\dfrac{dv}{dx} = 2x$.

Using the product rule, $\dfrac{dy}{dx} = v\dfrac{du}{dx} + u\dfrac{dv}{dx}$

$$= (x^2 - 5) \times 2 + (2x + 3) \times 2x$$

$$= 2(x^2 - 5 + 2x^2 + 3x)$$

$$= 2(3x^2 + 3x - 5)$$

NOTE

In this case you could have multiplied out the expression for y:

$$y = 2x^3 + 3x^2 - 10x - 15$$

$$\dfrac{dy}{dx} = 6x^2 + 6x - 10$$

$$= 2(3x^2 + 3x - 5).$$

EXAMPLE

Differentiate $y = 20x\,(x - 1)^6$

Solution

Let $u = 20x$ and $v = (x - 1)^6$

Then $\dfrac{du}{dx} = 20$, and $\dfrac{dv}{dx} = 6(x - 1)^5$ (using the chain rule).

Using the product rule, $\dfrac{dy}{dx} = v\dfrac{du}{dx} + u\dfrac{dv}{dx}$

$$= (x - 1)^6 \times 20 + 20x \times 6(x - 1)^5$$

$$= 20(x - 1)^5 \times (x - 1) + 20(x - 1)^5 \times 6x$$

$$= 20(x - 1)^5 \big[(x - 1) + 6x\big]$$

$$= 20(x - 1)^5 (7x - 1).$$

$(x - 1)^5$ is a common factor

The factorised result is the most useful form for the solution, as it allows you to find stationary points easily. You should always try to factorise your answer as much as possible. Once you have used the product rule, look for factors straight away and do not be tempted to multiply out.

The quotient rule

In the last section, you met a technique for differentiating the product of two functions. In this section you will see how to differentiate a function which is the quotient of two simpler functions.

As before, you start by identifying the simpler functions. For example, the function

$$y = \frac{3x+1}{x-2}$$

can be written as $y = \dfrac{u}{v}$ where $u = 3x + 1$ and $v = x - 2$. Using this

notation, $\dfrac{dy}{dx}$ is given by

$$\frac{dy}{dx} = \frac{v\dfrac{du}{dx} - u\dfrac{dv}{dx}}{v^2}.$$

This is called the *quotient rule*, and it is derived from first principles in the next section.

The quotient rule from first principles

A small increase, δx in x results in corresponding small increases δu, δv and δy in u, v and y. The new value of y is given by

$$y + \delta y = \frac{u + \delta u}{v + \delta v}$$

and since $y = \dfrac{u}{v}$, you can rearrange this to obtain an expression for δy in terms of u and v:

$$\delta y = \frac{u + \delta u}{v + \delta v} - \frac{u}{v}$$

$$= \frac{v(u + \delta u) - u(v + \delta v)}{v(v + \delta v)}$$

$$= \frac{uv + v\delta u - uv - u\delta v}{v(v + \delta v)}$$

$$= \frac{v\delta u - u\delta v}{v(v + \delta v)}$$

Dividing both sides by δx gives

$$\frac{\delta y}{\delta x} = \frac{v\dfrac{\delta u}{\delta x} - u\dfrac{\delta v}{\delta x}}{v(v + \delta v)}$$

In the limit as $\delta x \to 0$, this is written in the form you met on p116:

$$\frac{dy}{dx} = \frac{v\dfrac{du}{dx} - u\dfrac{dv}{dx}}{v^2}.$$

Activity

Verify that the quotient rule gives $\dfrac{dy}{dx}$ correctly when $u = x^{10}$ and $v = x^7$.

EXAMPLE Given that $y = \dfrac{3x+1}{x-2}$, find $\dfrac{dy}{dx}$ using the quotient rule.

Solution

(a) Letting $u = 3x + 1$ and $v = x - 2$ gives

$$\frac{du}{dx} = 3 \qquad \text{and} \qquad \frac{dv}{dx} = 1$$

Using the quotient rule, $\dfrac{dy}{dx} = \dfrac{v\dfrac{du}{dx} - u\dfrac{dv}{dx}}{v^2}$

$$= \frac{(x-2)3 - (3x+1)1}{(x-2)^2}$$

$$= \frac{3x - 6 - 3x - 1}{(x-2)^2}$$

$$= \frac{-7}{(x-2)^2}.$$

EXAMPLE Given that $y = \dfrac{x^2+1}{3x-1}$, find $\dfrac{dy}{dx}$ using the quotient rule.

Solution

Letting $u = x^2 + 1$ and $v = 3x - 1$ gives

$$\frac{du}{dx} = 2x \qquad \text{and} \qquad \frac{dv}{dx} = 3.$$

Using the quotient rule, $\dfrac{dy}{dx} = \dfrac{v\dfrac{du}{dx} - u\dfrac{dv}{dx}}{v^2}$

$$= \frac{(3x-1)2x - (x^2+1)3}{(3x-1)^2}$$

$$= \frac{6x^2 - 2x - 3x^2 - 3}{(3x-1)^2}$$

$$= \frac{3x^2 - 2x - 3}{(3x-1)^2}.$$

Exercise 4D

1. Differentiate the following functions using the product rule or the quotient rule.

(a) $y = (x^2 - 1)(x^3 + 3)$

(b) $y = x^5(3x^2 + 4x - 7)$

(c) $y = x^2(2x + 1)^4$

(d) $y = \dfrac{3}{2x - 1}$

(e) $y = \dfrac{2x}{3x - 1}$

(f) $y = \dfrac{x^3}{x^2 + 1}$

2. The graph of $y = \dfrac{x}{x-1}$ is shown below.

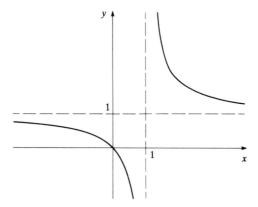

(i) Find $\dfrac{dy}{dx}$.

(ii) Find the gradient of the curve at $(0, 0)$, and the equation of the tangent at $(0, 0)$.

(iii) Find the gradient of the curve at $(2, 2)$, and the equation of the tangent at $(2, 2)$.

(iv) What can you deduce about the two tangents?

3. Given that $y = (x + 1)(x - 2)^2$,

(i) find $\dfrac{dy}{dx}$;

(ii) find any stationary points and determine their nature;

(iii) sketch the curve.

4. Given that $y = (2x - 1)^3 (x + 1)^3$,

(i) Find $\dfrac{dy}{dx}$ and factorise the expression you obtain;

(ii) Find the values of x for which $\dfrac{dy}{dx} = 0$, and determine the nature of the corresponding stationary points;

(iii) The graph of $y = (2x - 1)^3 (x + 1)^3$ is shown below. Write down the co-ordinates of P, Q and R.

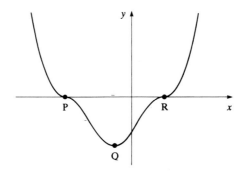

5. The graph of $y = \dfrac{2x}{\sqrt{x - 1}}$, which is undefined for $x < 0$ and $x = 1$, is shown below. P is a minimum point.

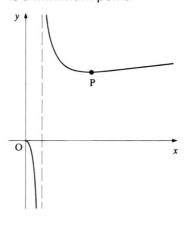

Exercise 4D continued

(i) Find $\dfrac{dy}{dx}$.

(ii) Find the gradient of the curve at (9, 9), and show that the equation of the normal at (9, 9) is $y = -4x + 45$.

(iii) Find the co-ordinates of P and verify that it is a minimum point.

(iv) Write down the equation of the tangent and the normal to the curve at P.

(v) Write down the point of intersection of

   (a) the normal found in (ii) and the tangent found in (iv), call it Q;

   (b) the normal found in (ii) and the normal found in (iv), call it R.

(vi) Show that the area of the triangle PQR is $\dfrac{441}{8}$.

6. Given that $y = \dfrac{x-3}{x-4}$,

(i) find $\dfrac{dy}{dx}$;

(ii) find the equation of the tangent to the curve at the point (6, 1.5);

(iii) find the equation of the normal to the curve at the point (5,2);

(iv) use your answer from (i) to deduce that the curve has no turning points, and sketch the graph.

7. Given that $y = \dfrac{x^2 - 2x - 5}{2x + 3}$;

(i) find $\dfrac{dy}{dx}$;

(ii) use your answer from (i) to find any stationary points of the curve;

(iii) classify each of the stationary points.

Differentiating an inverse function

Activity

What is the relationship between $\dfrac{dy}{dx}$ and $\dfrac{dx}{dy}$? Follow the steps below to help you to answer this question.

1. Differentiate $y = x^3$.

2. Rearrange $y = x^3$ in the form $x = f(y)$, and hence find $\dfrac{dx}{dy}$ as a function of y.

3. Write $\dfrac{dx}{dy}$ as a function of x.

4. Write down a relationship between $\dfrac{dy}{dx}$ and $\dfrac{dx}{dy}$.

5. Repeat steps 1–4 for other functions such as $y = 2x, y = x^2$ and $y = x^4$

6. Use your results to propose a general rule relating $\dfrac{dy}{dx}$ and $\dfrac{dx}{dy}$.

You may have proposed the general result $\dfrac{dy}{dx} = \dfrac{1}{\dfrac{dx}{dy}}$

If so, well done! The result looks algebraically obvious, but remember that

$\dfrac{dy}{dx}$ and $\dfrac{dx}{dy}$ are not fractions. The function $\dfrac{dy}{dx}$ is the rate of change of y

with x, and $\dfrac{dx}{dy}$ is the rate of change of x with y.

The geometrical interpretation of this result can be seen in figure 4.8 where, as an example, the line $y = \frac{1}{2} x$ is drawn first with the axes the normal way round, and then with them interchanged.

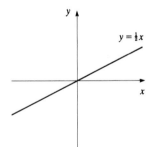

Figure 4.8

This demonstrates (but does not prove) that $\dfrac{dy}{dx} = \dfrac{1}{\dfrac{dx}{dy}}$

EXAMPLE

Given that $x = y^{\frac{1}{5}}$; find $\dfrac{dy}{dx}$

(i) by first finding $\dfrac{dx}{dy}$

(ii) by first making y the subject.

Solution

(i) $x = y^{\frac{1}{5}} \Rightarrow \dfrac{dx}{dy} = \dfrac{1}{5} \times y^{\frac{1}{5}-1}$

$$= \dfrac{1}{5} y^{-\frac{4}{5}}$$

$$= \dfrac{1}{5(y^{\frac{1}{5}})^4}$$

$$= \dfrac{1}{5x^4}.$$

Since $\dfrac{dy}{dx} = \dfrac{1}{\dfrac{dx}{dy}}$ it follows that $\dfrac{dy}{dx} = \dfrac{1}{\dfrac{1}{5x^4}}$

$$= 5x^4.$$

(ii) $x = y^{\frac{1}{5}} \Rightarrow y = x^5 \Rightarrow \dfrac{dy}{dx} = 5x^4$

Activity

Follow the steps below to *prove* that the rule

$$y = x^n \quad \Rightarrow \quad \dfrac{dy}{dx} = nx^{n-1}$$

is true for all positive rational numbers n. You use the fact that it is true for all positive integers.

Since n is a positive rational number it can be written as $n = \dfrac{p}{q}$ where p and q are positive integers.

Therefore $y = x^n = x^{p/q}$

Raising to the power q gives $\quad y^q = x^p = z$ (say).

(i) Find $\dfrac{dz}{dx}$ and $\dfrac{dz}{dy}$.

(ii) Use these results together with the chain rule and the relationship $\dfrac{dy}{dz} = \dfrac{1}{\dfrac{dz}{dy}}$ to show that $\dfrac{dy}{dx} = \dfrac{p}{q} x^{(\frac{p}{q}-1)}$.

Since $n = \dfrac{p}{q}$, this proves the result.

Integration

Earlier in this chapter you saw that the rule for differentiating powers of x could be extended from positive integer powers of x to all real powers of x. Since integration can be thought of as the reverse of differentiation, this suggests that the integration rule is not restricted just to positive integer powers of x.

In fact, it can be shown that the rule

$$\int kx^n \, dx = \frac{kx^{n+1}}{n+1} + c$$

may be extended to all real values of the power n, except -1.

Notice the very important exception, $n = -1$. This would give $n + 1 = 0$ on the bottom line of the answer, so the answer would be undefined. You will see in Chapter 5 that a different rule is applied when $n = -1$.

EXAMPLE Find the indefinite integral $\displaystyle\int \frac{4}{x^2} \, dx$

Solution

$$\int \frac{4}{x^2} \, dx = \int 4x^{-2} \, dx$$

$-2 + 1 = -1$

$$= \frac{4x^{-2+1}}{-1} + c$$

$$= -4x^{-1} + c$$

$$= -\frac{4}{x} + c$$

It is usual to put the final answer in the same form as the original function.

EXAMPLE Evaluate the definite integral $\displaystyle\int_4^9 x^{\frac{3}{2}} \, dx$

Solution

$$\int_4^9 x^{\frac{3}{2}} \, dx = \left[\frac{x^{\frac{5}{2}}}{\frac{5}{2}} \right]_4^9$$

$\frac{3}{2} + 1 = \frac{5}{2}$

$$= \frac{2}{5} \left[x^{\frac{5}{2}} \right]_4^9$$

$$= \frac{2}{5} (9^{\frac{5}{2}} - 4^{\frac{5}{2}})$$

$$= \frac{2}{5} (243 - 32)$$

$$= 84 \frac{2}{5}$$

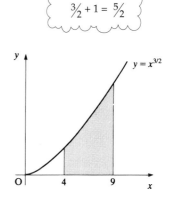

123

EXAMPLE   Find the area represented by $\int_1^2 \left(\dfrac{3}{x^4} - \dfrac{1}{x^2} + 4 \right) dx$.

Solution

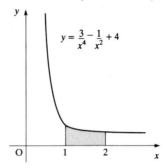

$$\int_1^2 \left(\frac{3}{x^4} - \frac{1}{x^2} + 4 \right) dx = \int_1^2 (3x^{-4} - x^{-2} + 4) \, dx$$

$$= \left[\frac{3x^{-3}}{-3} - \frac{x^{-1}}{-1} + 4x \right]_1^2$$

$$= \left[-\frac{1}{x^3} + \frac{1}{x} + 4x \right]_1^2$$

$$= \left(-\tfrac{1}{8} + \tfrac{1}{2} + 8 \right) - (-1 + 1 + 4)$$

$$= 4\tfrac{3}{8}.$$

The example below shows you how the extended rule of integration can help in solving differential equations.

EXAMPLE   Given that $\dfrac{dy}{dx} = \sqrt{x} + \dfrac{1}{x^2}$

(i)   find the general solution of the differential equation;

(ii)   find the equation of the curve with this gradient function which passes through (1, 5).

Solution

(i)

$$\frac{dy}{dx} = \sqrt{x} + \frac{1}{x^2}$$

$$= x^{1/2} + x^{-2}$$

$$\Rightarrow \quad y = \frac{x^{3/2}}{\frac{3}{2}} + \frac{x^{-1}}{-1} + c$$

$$= \frac{2}{3}x^{3/2} - \frac{1}{x} + c$$

(ii) Since the curve passes through $(1, 5)$

$$5 = \tfrac{2}{3} - 1 + c$$

$$c = 5\tfrac{1}{3}$$

$$\Rightarrow \qquad y = \tfrac{2}{3}x^{3/2} - \frac{1}{x} + 5\tfrac{1}{3}$$

Exercise 4E

1. In each of the following questions, find the indefinite integral. Remember to include the constant of integration.

(a) $\displaystyle\int 10x^{-4}\,dx$

(b) $\displaystyle\int (2x - 3x^{-4})\,dx$

(c) $\displaystyle\int (2 + x^3 + 5x^{-3})\,dx$

(d) $\displaystyle\int (6x^2 - 7x^{-2})\,dx$

(e) $\displaystyle\int 5x^{1/4}\,dx$

(f) $\displaystyle\int \frac{1}{x^4}\,dx$

(g) $\displaystyle\int \sqrt{x}\,dx$

(h) $\displaystyle\int \left(2x^4 - \frac{4}{x^2}\right)dx$

2. Evaluate the following definite integrals. Give answers as fractions or to 3 significant figures as appropriate.

(a) $\displaystyle\int_1^4 3x^{-2}\,dx$

(b) $\displaystyle\int_2^4 8x^{-3}\,dx$

(c) $\displaystyle\int_2^4 12x^{1/2}\,dx$

(d) $\displaystyle\int_{-3}^{-1} \frac{6}{x^3}\,dx$

(e) $\displaystyle\int_1^8 \left(\frac{x^2 + 3x + 4}{x^4}\right)dx$

(f) $\displaystyle\int_4^9 \left(\sqrt{x} - \frac{1}{\sqrt{x}}\right)dx$

3. The graph of $y = \sqrt{x} + \dfrac{1}{\sqrt{x}}$ for $x > 0$ is shown below.

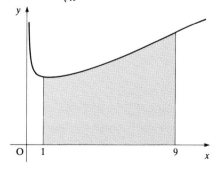

The shaded region is bounded by the curve, the x axis and the lines $x = 1$ and $x = 9$. Find its area.

Exercise 4E continued

4. The graph of $y = 4 - \dfrac{16}{x^2}$ is shown below, together with the line $y = 3$.

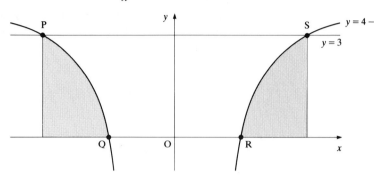

(i) Find the co-ordinates of the points P, Q, R and S.

(ii) Find the area of the shaded region.

5. (i) Find the point of intersection of the graphs of $y = \sqrt{x}$ and $y = \dfrac{128}{x^3}$.

(ii) Sketch, on the same axes, the graphs of $y = \sqrt{x}$ and $y = \dfrac{128}{x^3}$ for $0 \leq x \leq 6$.

(iii) Shade the region bounded by the curves, the x-axis and the line $x = 5$.

(iv) Find the area of the shaded region.

6. The graphs of $y = 1 - \dfrac{3}{x^2}$ and $y = -\dfrac{2}{x^3}$ are shown below.

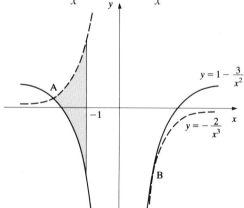

(i) Show that the x co-ordinates of the points where the curves meet are the roots of the equation

$$x^3 - 3x + 2 = 0$$

(ii) Hence show that the co-ordinates of A are $(-2, \frac{1}{4})$, and find the co-ordinates of point B.

(iii) Find the area of the shaded region.

7. Given that $\dfrac{dy}{dx} = \dfrac{2}{x^2} - 3$,

(i) find the general solution of the differential equation;

(ii) find the equation of the curve with this gradient function which passes through (2, 10).

8. Given that $\dfrac{dy}{dx} = \sqrt{x}$, find the general solution of this differential equation.

Find the equation of the curve with this gradient function which passes through (9, 20).

Integration by substitution

The graph of $y = \sqrt{(x-1)}$ is shown in figure 4.9.

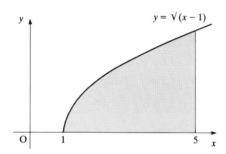

Figure 4.9

The shaded area is given by

$$\int_1^5 \sqrt{(x-1)}\, dx = \int_1^5 (x-1)^{\frac{1}{2}}\, dx.$$

You have not needed to find such an integral before, but you do know how to evaluate $\int_a^b u^{\frac{1}{2}}\, du$, so making the substitution $u = x-1$ will transform the integral into one that you can do.

When you make this substitution it means that you are now integrating with respect to a new variable, namely u. The limits of the integral, and the 'dx', must be written in terms of u.

The new limits are given by $\qquad x = 1 \qquad \Rightarrow \qquad u = 1 - 1 = 0$

$\qquad\qquad\qquad\qquad\quad$ and $\qquad x = 5 \qquad \Rightarrow \qquad u = 5 - 1 = 4.$

Since $u = x - 1$, $\dfrac{du}{dx} = 1.$

Even though $\dfrac{du}{dx}$ is not a fraction, it is usual to treat it as one in this situation (see Note below), and to write the next step as '$du = dx$'.

The integral now becomes

$$\int_{u=0}^{u=4} u^{\frac{1}{2}}\,du = \left[\frac{u^{\frac{3}{2}}}{\frac{3}{2}}\right]_{0}^{4}$$

$$= \left[\frac{2u^{\frac{3}{2}}}{3}\right]_{0}^{4}$$

$$= 5\tfrac{1}{3}$$

This method by integration is known as *integration by substitution*. It is a very powerful method which allows you to integrate many more functions. Since you are changing the variable from x to u, the method is also referred to as integration by change of variable.

NOTE

The last example included the statement '$du = dx$' Some mathematicians are reluctant to write such statements on the grounds that du and dx may only be used in the form $\dfrac{du}{dx}$, i.e. as a gradient. This is not in fact true; there is a well defined branch of mathematics which justifies such statements but it is well beyond the scope of this book. In the meantime it may help you to think of it as shorthand for 'in the limit as $\delta x \to 0$, $\dfrac{\delta u}{\delta x} \to 1$, and so $\delta u = \delta x$.'

EXAMPLE

Evaluate $\displaystyle\int_{1}^{3}(x+1)^3\,dx$ by making a suitable substitution.

Solution

Let $u = x + 1$.

Converting the limits: $x = 1 \;\Rightarrow\; u = 1 + 1 = 2$

$x = 3 \;\Rightarrow\; u = 3 + 1 = 4.$

Converting dx to du:

$$\frac{du}{dx} = 1 \qquad \Rightarrow \qquad du = dx.$$

$$\int_{1}^{3}(x+1)^3\,dx = \int_{2}^{4}u^3\,du$$

$$= \left[\frac{u^4}{4}\right]_{2}^{4}$$

$$= \frac{4^4}{4} - \frac{2^4}{4}$$

$$= 60.$$

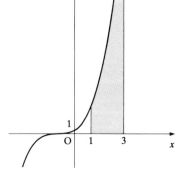

$y = (x+1)^3$

For Discussion

Can integration by substitution be described as the reverse of the chain rule?

EXAMPLE

Evaluate $\displaystyle\int_3^4 2x(x^2 - 4)^{\frac{1}{2}}\,dx$ by making a suitable substitution.

Solution

Notice that $2x$ is the derivative of the function in the brackets, $x^2 - 4$, and so $u = x^2 - 4$ is a natural substitution to try.

$$\text{This gives } \frac{du}{dx} = 2x \quad \Rightarrow \quad du = 2x\,dx.$$

Converting the limits:

$$x = 3 \quad \Rightarrow \quad u = 9 - 4 \ = 5.$$
$$x = 4 \quad \Rightarrow \quad u = 16 - 4 = 12.$$

So the integral becomes

$$\int_3^4 (x^2 - 4)^{\frac{1}{2}} 2x\,dx = \int_5^{12} u^{\frac{1}{2}}\,du$$

$$= \left[\frac{2u^{\frac{3}{2}}}{3} \right]_5^{12}$$

$$= 20.3 \qquad \text{(to 3 significant figures).}$$

In the last example there were two functions of x multiplied together, the second function being an expression in brackets raised to a power. The two functions are in this case related, since the first function, $2x$, is the derivative of the expression in brackets, $x^2 - 4$. It was this relationship that made the integration possible.

EXAMPLE

Find $\displaystyle\int x(x^2 + 2)^3\,dx$ by making an appropriate substitution.

Solution

Since this is an indefinite integral there are no limits to change, and the final answer will be a function of x.

Let $u = x^2 + 2$, then

$$\frac{du}{dx} = 2x \quad \Rightarrow \quad \text{‘}\tfrac{1}{2}du = x\,dx\text{’}$$

> You only have $x\,dx$ in the integral, not $2x\,dx$

So

$$\int x(x + 2)^3\,dx = \int (x + 2)^3 x\,dx$$

$$= \int u^3 \times \tfrac{1}{2}\,du$$

$$= \frac{u^4}{8} + c$$

$$= \frac{(x^2 + 2)^4}{8} + c.$$

Always remember, when finding an indefinite integral by substitution, to substitute back at the end. The original integral was in terms of x, so your final answer must be too.

EXAMPLE By making a suitable substitution, find $\int x\sqrt{(x-2)}\,dx$.

Solution

This question is not of the same type as the previous ones since x is not the derivative of $(x - 2)$. However, by making the substitution $u = x - 2$ you can still make the integral into one you can do.

Let $u = x - 2$, then

$$\frac{du}{dx} = 1 \quad \Rightarrow \quad du = dx.$$

There is also an x in the integral so you need to write down an expression for x in terms of u. Since $u = x - 2$ it follows that $x = u + 2$.

In the original integral you can now replace $\sqrt{(x - 2)}$ by $u^{\frac{1}{2}}$, dx by du, and x by $u + 2$.

$$\int x\sqrt{} \ (x-2)dx = \int (u+2)u^{\frac{1}{2}}\,du$$

$$= \int (u^{\frac{3}{2}} + 2u^{\frac{1}{2}})\,du$$

$$= \tfrac{2}{5}u^{\frac{5}{2}} + \tfrac{4}{3}u^{\frac{3}{2}} + c$$

Replacing u by $x-2$ and tidying up gives $\tfrac{2}{15}(3x+4)(x-2)^{\frac{3}{2}} + c$

Exercise 4F

1. Find the following indefinite integrals by making the suggested substitution. Remember to give your final answer in terms of x.

(a) $\int (x+1)^3\,dx,\ u = x + 1$

(b) $\int 2\sqrt{(2x-1)}\,dx,\ u = 2x - 1$

(c) $\int 3x^2(x^3+1)^7\,dx,\ u = x^3 + 1$

(d) $\int 2x(x^2+1)^5\,dx,\ u = x^2 + 1$

(e) $\int 3x^2(x^3-2)^4\,dx,\ u = x^3 - 2$

(f) $\int x\sqrt{(2x^2-5)}\,dx,\ u = 2x^2 - 5$

(g) $\int x\sqrt{(2x+1)}\,dx,\ u = 2x + 1$

(h) $\int \dfrac{x}{\sqrt{(x+9)}}\,dx,\ u = x + 9$

Exercise 4F continued

2. Evaluate each of the following definite integrals by using a suitable substitution. Give your answer to 3 significant figures where appropriate.

(a) $\int_{-1}^{4} (x-3)^4 \, dx$

(b) $\int_{0}^{3} (3x+2)^6 \, dx$

(c) $\int_{5}^{9} \sqrt{(x-5)} \, dx$

(d) $\int_{2}^{15} \sqrt[3]{(2x-3)} \, dx$

(e) $\int_{1}^{5} x^2 (x^3+1)^2 \, dx$

(f) $\int_{-1}^{2} 2x(x-3)^5 \, dx$

(g) $\int_{1}^{5} x\sqrt{(x-1)} \, dx$

3. The graph of $y = (x-2)^3$ is shown below.

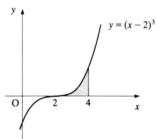

$y = (x-2)^3$

(i) Evaluate $\int_{2}^{4} (x-2)^3 \, dx$.

(ii) Without doing any calculations, state what do you think the value of $\int_{0}^{2} (x-2)^3 \, dx$ would be? Give reasons.

(iii) Confirm your answer by carrying out the integration.

4. The graph of $y = (x-1)^4 - 1$ is shown below.

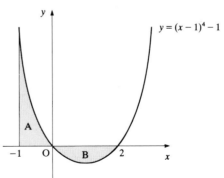

$y = (x-1)^4 - 1$

(i) Find the area of the shaded region A by evaluating $\int_{-1}^{0} \left((x-1)^4 - 1\right) dx$.

(ii) Find the area of the shaded region B by evaluating an appropriate integral.

(iii) Write down the total shaded area.

(iv) Why could you not just evaluate $\int_{-1}^{2} \left((x-1)^4 - 1\right) dx$ to find the total area?

Exercise 4F continued

5. Find the area of the shaded region for each of the following graphs.

(a)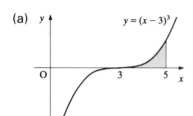

(b)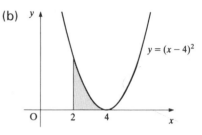

(c)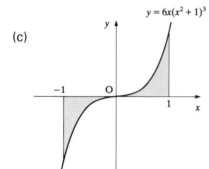

(d)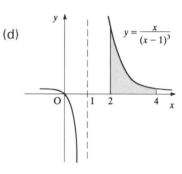

6.

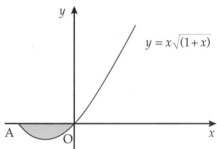

The sketch shows part of the graph of $y = x\sqrt{(1 + x)}$.
(i) Find the co−ordinates of point A and the range of values of x for which the function is defined.
(ii) Show that the shaded area is $\frac{4}{15}$. You may find the substitution $u = 1 + x$ useful.

[MEI]

7. (a) By substituting $u = 1 + x$ or otherwise, find

(i) $\int (1+x)^3 \, dx$

(ii) $\int_{-1}^{1} x(1+x)^3 \, dx.$

(b) By substituting $t = 1 + x^2$ or otherwise, evaluate $\int_{0}^{1} x\sqrt{(1+x^2)} \, dx.$

[MEI]

Exercise 4F continued

8. Sketch the graph of $y = \sqrt{(4+x^2)}$ showing any asymptotes.

Find the area contained between the curve and the x axis for $0 \le x \le 2$.

[MEI]

KEY POINTS

- $y = kx^n \implies \dfrac{dy}{dx} = knx^{n-1}$ where k and n are real constants

- Chain rule: $\dfrac{dy}{dx} = \dfrac{dy}{du} \times \dfrac{du}{dx}$

- Product rule (for $y=uv$): $\dfrac{dy}{dx} = v\dfrac{du}{dx} + u\dfrac{dv}{dx}$

- Quotient rule (for $y = \dfrac{u}{v}$): $\dfrac{dy}{dx} = \dfrac{v\dfrac{du}{dx} - u\dfrac{dv}{dx}}{v^2}$.

- **Stationary points:** $\dfrac{dy}{dx} = 0$

  * If $\dfrac{d^2y}{dx^2} < 0$, the point is a maximum.

  * If $\dfrac{d^2y}{dx^2} > 0$, the point is a minimum.

  * If $\dfrac{d^2y}{dx^2} = 0$, the point could be a maximum, a minimum or a point of inflection.

- $\dfrac{dy}{dx} = \dfrac{1}{\dfrac{dx}{dy}}$

- $\displaystyle\int kx^n \, dx = \dfrac{kx^{n+1}}{n+1} + c$ where k and n are constants but $n \ne -1$

- Substitution is often used to change a non-standard integral into a standard one.

5

Natural logarithms and exponentials

Normally speaking it may be said that the forces of a capitalist society, if left unchecked, tend to make the rich richer and the poor poorer and thus increase the gap between them.

Jawaharlal Nehru

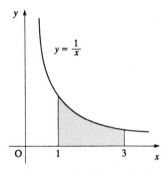

The shaded region on the graph above is bounded by the x axis, the lines $x = 1$ and $x = 3$, and the curve $y = \frac{1}{x}$. The area of this region may be represented by $\int_1^3 \frac{1}{x} \, dx$. How do you evaluate this integral?

You have probably decided that you are unable to evaluate this integral: the power of x is -1, and this is the one value of n for which you cannot apply the rule

$$\int kx^n \, dx = \frac{kx^{n+1}}{n+1} + c$$

If you were to apply this rule to $\int x^{-1} \, dx$, you would obtain $\frac{x^0}{0} + c$ and division by zero is undefined.

However, the area in the diagram clearly has a definite value, and so we need to find ways to express and calculate it.

Investigation

Estimate, using numerical integration (e.g. the trapezium rule), the areas represented by

(i) $\int_1^3 \frac{1}{x} \, dx$ (ii) $\int_1^2 \frac{1}{x} \, dx$ (iii) $\int_1^6 \frac{1}{x} \, dx$

What relationship can you see between your answers?

A new function

The area under the curve $y = \frac{1}{x}$ between $x = 1$ and $x = a$, that is $\int_1^a \frac{1}{x} dx$, depends on the value a. For every value of a (greater than 1) there is a definite value of the area. Consequently, the area is a function of a.

To investigate this function you need to give it a name, say L, so that L(a) is the area from 1 to a and L(x) is the area from 1 to x. Then look at the properties of L(x) to see if its behaviour is like that of any other function with which you are familiar.

The investigation you have just done should have suggested to you that

$$\int_1^3 \frac{1}{x} dx + \int_1^2 \frac{1}{x} dx = \int_1^6 \frac{1}{x} dx.$$

This can now be written as

$$L(3) + L(2) = L(6)$$

This suggests a possible law, that

$$L(a) + L(b) = L(ab).$$

At this stage this is just a conjecture, based on one particular example. To prove it, you need to take the general case and this is done in the activity below. (At first reading you may prefer to leave the activity, accepting that the result can be proved.)

Activity

Prove that L(a) + L(b) = L(ab), by following the steps below.

(i)   Explain, with the aid of a diagram, why

$$L(a) + \int_a^{ab} \frac{1}{x} dx = L(ab)$$

(ii) By substituting $x = az$, show that

$$\int_a^{ab} \frac{1}{x} dx = \int_1^b \frac{1}{z} dz$$

Explain why $\int_1^b \frac{1}{z} dz = L(b)$

(iii) Use the results from parts (i) and (ii) to show that

$$L(a) + L(b) = L(ab)$$

What function has this property? Look back to Chapter 1, and you will see that for all logarithms

$$\log(a) + \log(b) = \log(ab)$$

Could it be that this is a logarithmic function?

Activity

Satisfy yourself that the function has the following properties of logarithms

(i) $L(1) = 0$ (ii) $L(a) - L(b) = L\left(\frac{a}{b}\right)$ (iii) $L(a^n) = nL(a)$

The base of the logarithm function $L(x)$

Having accepted that $L(x)$ is indeed a logarithmic function, the remaining problem is to find the base of the logarithm. By convention this is denoted by the letter e. A further property of logarithms is that for any base p

$$\log_p p = 1 \qquad (p > 1).$$

So to find the base e, you need to find the point such that the area, $L(e)$ under the graph, is 1 (figure 5.1).

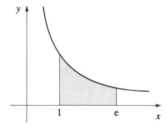

Figure 5.1

You have already estimated the value of $L(2)$ to be about 0.7 and that of $L(3)$ to be about 1.1 so clearly the value of e is between 2 and 3.

Activity

(You will need a calculator with an area-finding facility, or other suitable technology, to do this. If you do not have this, read on.)

Use the fact that $\int_1^e \frac{1}{x}\,dx = 1$ to find the value of e, knowing that it lies between 2 and 3, to 2 decimal places.

The value of e is given to 9 decimal places in the key points on page 157. Like π, e is a number which occurs naturally within mathematics. It is irrational: when written as a decimal, it never terminates and has no recurring pattern.

The function $L(x)$ is thus the logarithm of x to the base e, $\log_e x$. This is often called the natural logarithm of x, and written as $\ln x$.

Values of x between 0 and 1

So far it has been assumed that the value of x within $\ln x$ is greater than 1. As an example of a value of x between 0 and 1, look at $\ln \frac{1}{2}$.

Since $\quad \ln \left(\frac{a}{b}\right) = \ln a - \ln b$

$\Rightarrow \qquad \ln \left(\frac{1}{2}\right) = \ln 1 - \ln 2$

$\qquad\qquad = - \ln 2 \qquad$ (since $\ln 1 = 0$).

In the same way, you can show that for any value of x between 0 and 1, the value of $\ln x$ is negative.

When the value of x is very close to zero, the value of $\ln x$ is a large negative number:

$$\ln \left(\frac{1}{1000}\right) = - \ln 1000 = -6.9$$

$$\ln \left(\frac{1}{1000\,000}\right) = - \ln 1\,000\,000 = -13.8$$

In the limit as $x \to 0$, $\ln x \to - \infty$ (for positive values of x).

The natural logarithm function

The graph of the natural logarithm function (shown in figure 5.2) has the characteristic shape of all logarithm functions, and like other such functions it is only defined for $x > 0$. The value of $\ln x$ increases without limit, but ever more slowly: it has been described as 'the slowest way to get to infinity'.

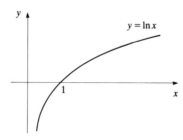

Figure 5.2

NOTE

Logarithms were discovered independently by John Napier (1550–1617), who lived at Merchiston Castle in Edinburgh, and Jolst Bürgi (1552–1632) from Switzerland. It is generally believed that Napier had the idea first, and so he is credited with their discovery. Natural logarithms are also called Naperian logarithms but there is no basis for this since Napier's logarithms were definitely not the same as natural logarithms. Napier was deeply involved in the political and religious events of his day and mathematics and science were little more than hobbies for him. He was a man of remarkable ingenuity and imagination and also drew plans for war chariots that look very like modern tanks, and for submarines.

The exponential function

The graph of the inverse of the natural logarithm function is found by reflecting $y = \log_e x$ in the line $y = x$, and its equation by interchanging x and y to get $x = \log_e y$.

Making y the subject of $x = \log_e y$ using the theory of logarithms developed in chapter 1 of this book, you obtain $e^x = y$ or $y = e^x$. The graphs of the logarithm function and its inverse are shown in figure 5.3.

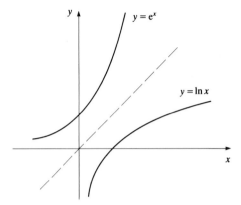

Figure 5.3

Since the function e^x is the inverse of $\ln x$, and vice versa, it follows that

$$e^{(\ln x)} = x \qquad \text{and} \qquad \ln(e^x) = x.$$

The function e^x is often called the *exponential function*. In fact, any function of the form a^x is exponential. Figure 5.4 shows several exponential curves.

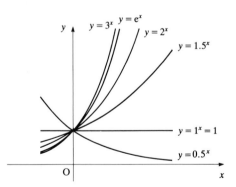

Figure 5.4

The exponential function $y = e^x$ increases at an ever-increasing rate. This is described as exponential growth.

By contrast, the graph of $y = e^{-x}$ approaches the x axis ever more slowly as x increases: this is called exponential decay (figure 5.5).

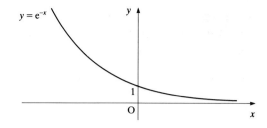

Figure 5.5

EXAMPLE

The number, N, of insects in a colony is given by $N = 2000e^{0.1t}$ where t is the number of days after observations have begun.
(i)   Sketch the graph of N against t.
(ii) What is the population of the colony after 20 days?
(iii) How long does it take the colony to reach a population of 10000?

Solution

(i)

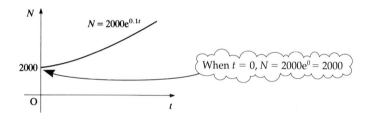

(ii) When $t = 20$, $N = 2000e^{0.1 \times 20}$

$$= 14778$$

The population is 14 778 insects.

(iii) When $N = 10000$, $10000 = 2000\, e^{0.1t}$

$$5 = e^{0.1t}$$

Taking natural logarithms of both sides,

$$\ln 5 = \ln(e^{0.1t})$$

Remember
$\ln(e^x) = x$

$$= 0.1t$$

and so $t = 10 \ln 5$

$$t = 16.09\ldots$$

It takes just over 16 days for the population to reach 10,000.

EXAMPLE

The radioactive mass, M g in a lump of material is given by
$M = 25e^{-0.0012t}$ where t is the time in seconds since the first observation.
(i) Sketch the graphs of M against t.
(ii) What is the initial size of the mass?
(iii) What is the mass after 1 hour?
(iv) The half-life of a radioactive substance is the time it takes to decay to half of its mass. What is the half-life of this material?

Solution

(i)

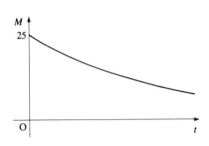

(ii) When $t = 0$, $\quad M = 25\,e^0$
$$= 25.$$

The initial mass is 25 g.

(iii) After one hour, $\quad t = 3600$
$$M = 25e^{-0.0012 \times 3600}$$

The mass after one hour is 0.33 g (to 2 decimal places).

(iv) The initial mass is 25 g, so after one half-life,
$$M = \tfrac{1}{2} \times 25 = 12.5\text{g}$$

At this point the value of t is given by
$$12.5 = 25e^{-0.0012t}$$

Dividing both sides by 25 gives
$$0.5 = e^{-0.0012t}$$

Taking logarithms of both sides:
$$\ln 0.5 = \ln e^{-0.0012t}$$
$$= -0.0012t$$
$$\Rightarrow \quad t = \frac{\ln 0.5}{-0.0012}$$
$$= 557.6 \text{ (to 1 decimal place).}$$

The half-life is 577.6 seconds. (This is just under 10 minutes, so the substance is highly radioactive).

EXAMPLE   Make p the subject of $\ln(p) - \ln(1-p) = t$

Solution

$$\ln\left(\frac{p}{1-p}\right) = t \qquad \text{using } \log a - \log b = \log\left(\frac{a}{b}\right)$$

Writing both sides as powers of e gives

$$e^{\ln\left(\frac{p}{1-p}\right)} = e^t$$

$$\Rightarrow \quad \frac{p}{1-p} = e^t \qquad \text{Remember } e^{\ln x} = x$$

$$p = e^t(1-p)$$

$$p = e^t - pe^t$$

$$p + pe^t = e^t$$

$$p(1+e^t) = e^t$$

$$p = \frac{e^t}{1+e^t}$$

Exercise 5A

1. Make x the subject of $\ln x - \ln x_0 = kt.$

2. Make t the subject of $s = s_0 e^{-kt}$

3. Make p the subject of $\ln\left(\dfrac{p}{25}\right) = -0.02t$

4. Make x the subject of $y - 5 = (y_0 - 5)e^x$

5. A colony of humans settles on a previously uninhabited planet. After t years, their population, P, is given by

$$P = 100e^{0.05t}$$

(i)   Sketch the graph of P against t.
(ii) How many settlers land on the planet initially?
(iii) What is the population after 50 years?
(iv) How long does it take the population to reach 1 million?

Exercise 5A continued

6. Ela sits on a swing. Her father pulls it back and then releases it. The swing returns to its maximum backwards displacement once every 5 seconds, but the maximum displacement, $\theta°$, becomes progressively smaller because of friction. At time t seconds, θ is given by

$$\theta = 25e^{-0.03t} \quad (t = 0, 5, 10, 15, \ldots)$$

(i)   Plot the values of θ for $0 \le t \le 30$ on graph paper.
(ii) To what angle did Ela's father pull the swing?
(iii) What is the value of θ after one minute?
(iv) After how many swings is the angle θ less than $1°$?

7. Alexander lives 800 metres from school. One morning he sets out at 8.00 a.m. and t minutes later the distance s m, which he has walked is given by
$$s = 800\,(1 - e^{-0.1t})$$

(i) Sketch the graph of s against t.
(ii) How far has Alexander walked by 8.15 a.m.?
(iii) What time is it when Alexander is half way to school?
(iv) When does Alexander get to school?

8. A parachutist jumps out of an aircraft and some time later opens the parachute. His speed at time t seconds from when the parachute opens is $v\,\text{ms}^{-1}$. It is given by

$$v = 8 + 22\,e^{-0.07t}$$

(i)   Sketch the graph of v against t.
(ii) State the speed of the parachutist when the parachute opens, and the final speed that he would attain if he jumped from a very great height.
(iii) Find the value of v as the parachutist lands, 60 seconds later.
(iv) Find the value of t when the parachutist is travelling at $20\,\text{ms}^{-1}$.

Differentiating natural logarithms and exponentials

Since the integral of $\frac{1}{x}$ is $\ln x$, it follows that the differential of $\ln x$ is $\frac{1}{x}$.

So $y = \ln x \quad \Rightarrow \quad \dfrac{dy}{dx} = \dfrac{1}{x}.$

The differential of the inverse function, $y = e^x$, may be found by interchanging y and x
$$x = \ln y$$

$$\Rightarrow \qquad \frac{dx}{dy} = \frac{1}{y}$$

$$\Rightarrow \qquad \frac{dy}{dx} = \frac{1}{\dfrac{dx}{dy}} = y = e^x$$

Therefore $\qquad \dfrac{d}{dx} e^x = e^x.$

The differential of e^x is itself e^x. This may at first seem rather surprising.

It follows that the integral of e^x is also e^x.

$$\int e^x \, dx = e^x + c$$

The work in the preceding pages may be summarised as in the following table.

Differentiation	Integration
$y \longrightarrow \dfrac{dy}{dx}$	$y \longrightarrow \displaystyle\int y \, dx$
$\ln x \longrightarrow \dfrac{1}{x}$	$\dfrac{1}{x} \longrightarrow \ln x + c$
$e^x \longrightarrow e^x$	$e^x \longrightarrow e^x + c$

These results allow you to extend very considerably the range of functions which you are able to differentiate and integrate.

EXAMPLE

Differentiate $y = e^{5x}$.

Solution

Make the substitution $u = 5x$ to give $y = e^u$.

Now $\dfrac{dy}{du} = e^u = e^{5x}$ $\qquad$ and $\qquad \dfrac{du}{dx} = 5$

By the chain rule, $\qquad \dfrac{dy}{dx} = \dfrac{dy}{du} \times \dfrac{du}{dx}$

$$= e^{5x} \times 5$$

$$= 5e^{5x}.$$

This result can be generalised as follows:

$$y = e^{ax} \quad \Rightarrow \quad \frac{dy}{dx} = ae^{ax} \quad \text{where } a \text{ is any constant.}$$

This is an important standard result, and you would normally use it automatically, without recourse to the chain rule.

EXAMPLE Differentiate $y = \dfrac{4}{e^{2x}}$.

Solution

$$y = \frac{4}{e^{2x}} = 4e^{-2x}$$

$$\Rightarrow \quad \frac{dy}{dx} = 4 \times (-2e^{-2x})$$

$$= -8e^{-2x}$$

EXAMPLE Differentiate $y = 3e^{(x^2+1)}$.

Solution

Let $u = x^2 + 1$, then $y = 3e^u$.

$$\Rightarrow \quad \frac{dy}{du} = 3e^u = 3e^{(x^2+1)} \quad \text{and} \quad \frac{du}{dx} = 2x.$$

By the chain rule, $\quad \dfrac{dy}{dx} = \dfrac{dy}{du} \times \dfrac{du}{dx}$

$$= 3e^{(x^2+1)} \times 2x$$

$$= 6xe^{(x^2+1)}.$$

EXAMPLE Differentiate the following functions.

(a) $y = 2\ln x$ (b) $y = \ln(3x)$

Solution

(a) $\dfrac{dy}{dx} = 2 \times \dfrac{1}{x}$

$$= \frac{2}{x}.$$

(b) Let $u = 3x$, then $y = \ln u$

$$\Rightarrow \quad \frac{dy}{du} = \frac{1}{u} = \frac{1}{3x} \quad \text{and} \quad \frac{du}{dx} = 3$$

By the chain rule:

$$\frac{dy}{dx} = \frac{dy}{du} \times \frac{du}{dx}$$

$$= \frac{1}{3x} \times 3$$

$$= \frac{1}{x}.$$

NOTE

An alternative solution to part (b) is:

$$y = \ln(3x) = \ln 3 + \ln x \qquad \Rightarrow \qquad \frac{dy}{dx} = 0 + \frac{1}{x} = \frac{1}{x}.$$

For Discussion

The gradient function found in part (b) above for $y = \ln(3x)$ is the same as that for $y = \ln(x)$. What does this tell you about the shapes of the two curves?

EXAMPLE

Differentiate the following functions.

(a) $y = \ln(x^4)$ (b) $y = \ln(x^2 + 1)$

Solution

(a) By the properties of logarithms:

$$y = \ln(x^4)$$

$$= 4\ln(x)$$

$$\Rightarrow \qquad \frac{dy}{dx} = \frac{4}{x}.$$

(b) Let $u = x^2 + 1$, then $y = \ln u$.

$$\Rightarrow \qquad \frac{dy}{du} = \frac{1}{u} = \frac{1}{x^2 + 1} \qquad \text{and} \qquad \frac{du}{dx} = 2x$$

By the chain rule,

$$\frac{dy}{dx} = \frac{dy}{du} \times \frac{du}{dx}$$

$$= \frac{1}{x^2 + 1} \times 2x$$

$$= \frac{2x}{x^2 + 1}.$$

If you need to differentiate functions similar to those in the above examples, follow exactly the same steps. The results could be generalised as follows.

$$y = a\ln x \qquad \Rightarrow \qquad \frac{dy}{dx} = \frac{a}{x}$$

$$y = \ln(ax) \qquad \Rightarrow \qquad \frac{dy}{dx} = \frac{1}{x}$$

$$y = \ln(f(x)) \qquad \Rightarrow \qquad \frac{dy}{dx} = \frac{f'(x)}{f(x)}$$

EXAMPLE Differentiate $y = \dfrac{\ln x}{x}$

Solution

Here y is of the form $\dfrac{u}{v}$ where $u = \ln x$ and $v = x$.

$$\Rightarrow \qquad \frac{du}{dx} = \frac{1}{x} \quad \text{and} \quad \frac{dv}{dx} = 1.$$

By the quotient rule, $\dfrac{dy}{dx} = \dfrac{v\dfrac{du}{dx} - u\dfrac{dv}{dx}}{v^2}$

$$= \frac{x \times \dfrac{1}{x} - 1 \times \ln x}{x^2}$$

$$= \frac{1 - \ln x}{x^2}$$

Exercise 5B

1. Differentiate the following functions.

(a) $y = 3\ln x$ (b) $y = \ln(4x)$ (c) $y = \ln(x^2)$

(d) $y = \ln(x^2 + 1)$ (e) $y = \ln\left(\dfrac{1}{x}\right)$ (f) $y = x\ln x$

(g) $y = x^2\ln(4x)$ (h) $y = \ln\left(\dfrac{x+1}{x}\right)$ (i) $y = \ln\sqrt{(x^2 - 1)}$

(j) $y = \dfrac{\ln x}{x^2}$

2. Differentiate the following functions.

(a) $y = 3e^x$ (b) $y = e^{2x}$

(c) $y = e^{x^2}$ (d) $y = e^{(x+1)^2}$

(e) $y = xe^{4x}$ (f) $y = 2x^3e^{-x}$

(g) $y = \dfrac{x}{e^x}$ (h) $y = (e^{2x} + 1)^3$

3. Knowing how much rain has fallen in a river basin, hydrologists are often able to give forecasts of what will happen to a river level over the next few hours. In one case it is predicted that the height h, in metres, of a river above its normal level during the next three hours will be $0.12e^{0.9t}$, where t is the time elapsed, in hours, after the prediction.

   (i) Find $\dfrac{dh}{dt}$, the rate at which the river is rising.

   (ii) At what rate will the river be rising after $0, 1, 2$ and 3 hours?

4. The graph of $y = xe^x$ is shown below.

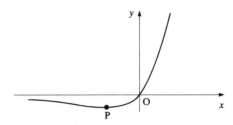

   (i) Find $\dfrac{dy}{dx}$ and $\dfrac{d^2y}{dx^2}$.

   (ii) Find the co-ordinates of the minimum point P.

5. The graph of $y = x\ln(x^2)$ is shown below.

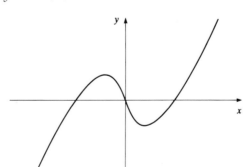

   (i) Find $\dfrac{dy}{dx}$ and $\dfrac{d^2y}{dx^2}$.

   (ii) Find the co-ordinates of any stationary points.

6. Given that $y = \dfrac{e^x}{x}$

   (i) find $\dfrac{dy}{dx}$

   (ii) find the co-ordinates of any stationary points on the curve of the function.

   (iii) sketch the curve.

Exercise 5B continued

7. (i) Differentiate $\ln x$ and $x \ln x$ with respect to x.

The sketch shows the graph of $y = x \ln x$ for $0 \leqslant x \leqslant 3$.

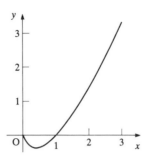

(ii) Show that the curve has a stationary point $\left(\dfrac{1}{e}, -\dfrac{1}{e} \right)$

[MEI]

8.

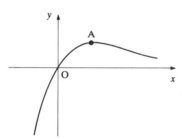

The diagram shows the graph of $y = xe^{-x}$.

(i) Differentiate xe^{-x}.

(ii) Find the co-ordinates of the point A, the maximum point on the curve.

[MEI]

Integrals involving the exponential function

Since you know that
$$\frac{d}{dx}(e^{ax}) = ae^{ax},$$

you can see that
$$\int e^{ax}\, dx = \frac{1}{a}e^{ax} + c.$$

This increases the number of functions which you are able to integrate, as in the following example.

EXAMPLE

Find the following integrals.

(a) $\int e^{2x} \, dx$ (b) $\int_1^5 6e^{3x} \, dx$

Solution

(a) $\int e^{2x} \, dx = \frac{1}{2} e^{2x} + c.$

(b) $\int_1^5 6e^{3x} \, dx = \left[\frac{6e^{3x}}{3} \right]_1^5$

$= \left[2e^{3x} \right]_1^5$

$= 2(e^{15} - e^3)$

$= 6.54 \times 10^6$ (to 3 significant figures).

EXAMPLE

By making a suitable substitution, find $\int_0^4 2xe^{x^2} \, dx$

Solution

$$\int_0^4 2xe^{x^2} \, dx = \int_0^4 e^{x^2} 2x \, dx.$$

Since $2x$ is the derivative of x^2, let $u = x^2$.

$$\frac{du}{dx} = 2x \quad \Rightarrow \quad du = 2x \, dx$$

The new limits are given by $x = 0 \quad \Rightarrow \quad u = 0$
and $x = 4 \quad \Rightarrow \quad u = 16.$

The integral can now be written as

$$\int_0^{16} e^u \, du = \left[e^u \right]_0^{16}$$

$= e^{16} - e^0$

$= 8.89 \times 10^6$ to 3 significant figures.

Integrals involving the natural logarithm function

You have already seen that

$$\int \frac{1}{x} \, dx = \ln x + c.$$

There are many other integrals that can be reduced to this form either by rearrangement or by substitution.

EXAMPLE Evaluate $\displaystyle\int_{2}^{5} \frac{1}{2x}\,dx$

Solution

$$\frac{1}{2}\int_{2}^{5}\frac{1}{x}\,dx = \frac{1}{2}\big[\ln x\big]_{2}^{5}$$

$$= \tfrac{1}{2}(\ln 5 - \ln 2)$$

$$= 0.458 \qquad \text{(to 3 significant figures)}.$$

In this example the $\frac{1}{2}$ was taken outside the integral, allowing the standard result for $\frac{1}{x}$ to be used. It is not always possible to do this, and in the following example a substitution is necessary.

EXAMPLE Evaluate $\displaystyle\int_{1}^{5} \frac{2x}{x^2+3}\,dx$

Solution

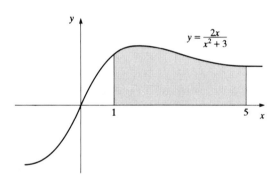

In this case, substitute $u = x^2 + 3$, so that

$$\frac{du}{dx} = 2x \qquad \Rightarrow \qquad du = 2x\,dx$$

The new limits are given by $\quad x = 1 \quad \Rightarrow \quad u = 4$

$$\text{and} \qquad x = 5 \quad \Rightarrow \quad u = 28.$$

$$\int_{1}^{5}\frac{2x}{x^2+3}\,dx = \int_{4}^{28}\frac{1}{u}\,du$$

$$= \big[\ln u\big]_{4}^{28}$$

$$= \ln 28 - \ln 4$$

$$= 1.95 \qquad \text{(to 3 significant figures)}.$$

The last example is of the form $\int \dfrac{f'(x)}{f(x)}\,dx$ where $f(x) = x^2 + 3$. In such

cases the substitution $u = f(x)$ transforms the integral into $\int \dfrac{1}{u}\,du$. The

answer is then $\ln u + c$ or $\ln(f(x)) + c$ (assuming that $u = f(x)$ is positive). This result may be stated as a working rule:

If you obtain the top line when you differentiate the bottom line, the integral is the natural logarithm of the bottom line.

EXAMPLE   Evaluate $\displaystyle\int_1^2 \dfrac{5x^4 + 2x}{x^5 + x^2 + 4}\,dx$

Solution

You can work this out by substituting $u = x^5 + x^2 + 4$, but since differentiating the bottom line gives the top line, you could apply the above rule and just write

$$\int_1^2 \dfrac{5x^4 + 2x}{x^5 + x^2 + 4}\,dx = \left[\ln(x^5 + x^2 + 4)\right]_1^2$$

$$= \ln 40 - \ln 6 = 1.90 \quad \text{(to 2 significant figures)}.$$

In the next example some adjustment is needed to get the top line into the required form.

EXAMPLE   Evaluate $\displaystyle\int_0^1 \dfrac{x^5}{x^6 + 7}\,dx$

Solution

The differential of $x^6 + 7$ is $6x^5$, so the integral is rewritten as $\dfrac{1}{6}\displaystyle\int_0^1 \dfrac{6x^5}{x^6 + 7}\,dx$

Integrating this gives $\dfrac{1}{6}\left[\ln(x^6 + 7)\right]_0^1$ or 0.022 (to 2 significant figures).

Extending the domain for logarithmic integrals

The use of $\displaystyle\int \dfrac{1}{x}\,dx = \ln x + c$ has so far been restricted to cases where $x > 0$,

since logarithms are undefined for negative numbers.

Similarly, for $\int \dfrac{f'(x)}{f(x)} \, dx = \ln f(x) + c$ it has been required that $f(x) > 0$.

Look however at the area between $-b$ and $-a$ on the left hand branch of the curve $y = \frac{1}{x}$ (figure 5.6). You can see that it is a real area, and that it must be possible to evaluate it.

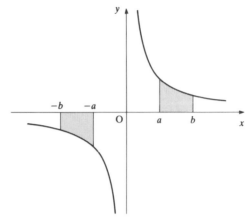

Figure 5.6

Activity

1. What can you say about the two shaded areas?
2. Try to prove your answer to 1 before reading on.

Proof

Let $A = \int_{-b}^{-a} \dfrac{1}{x} \, dx$.

Substituting $u = -x$ gives new limits: $x = -b \quad \Rightarrow \quad u = b$

$\qquad\qquad\qquad\qquad\qquad\qquad\qquad\qquad\quad x = -a \quad \Rightarrow \quad u = a.$

$$\frac{du}{dx} = -1 \;\Rightarrow\; dx = -du$$

So the integral becomes

$$A = \int_{b}^{a} \frac{1}{-u} (-du)$$

$$= \int_{b}^{a} \frac{1}{u} \, du$$

$$= \big[\ln a - \ln b \big]$$

So the area has the same size as that obtained if no notice is taken of the fact that the limits a and b have minus signs. However it has the opposite sign, as you would expect because the area is below the axis.

Consequently the restriction that $x > 0$ may be dropped, and the integral is written

$$\int \frac{1}{x}dx = \ln|x| + c.$$

Similarly, $$\int \frac{f'(x)}{f(x)}dx = \ln|f(x)| + c.$$

EXAMPLE   Find the value of $$\int_5^7 \frac{1}{4-x}dx$$

Solution

To make the top line into the differential of the bottom line, you write the integral as

$$-\int_5^7 \frac{-1}{4-x}dx = -\Big[\ln|4-x|\Big]_5^7$$

$$= -\Big[\big(\ln|-3|\big) - \big(\ln|-1|\big)\Big]$$

$$= -\big[\ln 3 - \ln 1\big]$$

$$= -1.10 \quad \text{(to 3 significant figures)}.$$

Caution

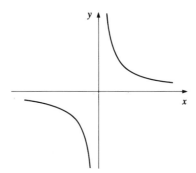

Figure 5.7

Since the curve is not defined at a discontinuity such as the one at $x = 0$ on the graph of $y = \frac{1}{x}$ (see figure 5.7), it is not possible to integrate across one.

Consequently in the integral $\int_p^q \frac{1}{x}dx$, both the limits p and q must have the same sign, either $+$ or $-$. The integral is invalid otherwise.

Exercise 5C

1. Find the following indefinite integrals.

(a) $\displaystyle\int \frac{3}{x}\,dx$

(b) $\displaystyle\int \frac{1}{4x}\,dx$

(c) $\displaystyle\int \frac{1}{x-5}\,dx$

(d) $\displaystyle\int \frac{1}{2x-9}\,dx$

(e) $\displaystyle\int \frac{2x}{x^2+1}\,dx$

(f) $\displaystyle\int \frac{2x+3}{3x^2+9x-1}\,dx$

2. Find the following indefinite integrals.

(a) $\displaystyle\int e^{3x}\,dx$

(b) $\displaystyle\int e^{-4x}\,dx$

(c) $\displaystyle\int e^{-x/3}\,dx$

(d) $\displaystyle\int 12x^2 e^{x^3}\,dx$

(e) $\displaystyle\int \frac{10}{e^{5x}}\,dx$

(f) $\displaystyle\int \frac{e^{3x}+4}{e^{2x}}\,dx$

3. Find the following definite integrals. Where appropriate give your answers to 3 significant figures.

(a) $\displaystyle\int_0^4 4e^{2x}\,dx$

(b) $\displaystyle\int_1^3 \frac{4}{2x+1}\,dx$

(c) $\displaystyle\int_2^6 2xe^{-x^2}\,dx$

(d) $\displaystyle\int_{-1}^1 (e^x + e^{-x})\,dx$

(e) $\displaystyle\int_{-2}^1 e^{3x-2}\,dx$

(f) $\displaystyle\int_2^4 \frac{x+2}{x^2+4x-3}\,dx$

4. The graph of $y = xe^{x^2}$ is shown below.

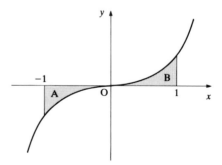

(i)   Find the area of region A.

(ii) Find the area of region B.

(iii) Hence write down the total area of the shaded region.

5. The graph of $y = xe^{-x^2}$ is shown below.

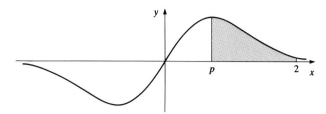

(i) Find $\dfrac{dy}{dx}$ using the product rule.

(ii) Find the x co-ordinate, p, of the maximum point. (You do not need to prove that it corresponds to a maximum.)

(iii) Use your answer from (ii) to find the area of the shaded region.

6. The graph of $y = \dfrac{x+2}{x^2 + 4x + 3}$ is shown below.

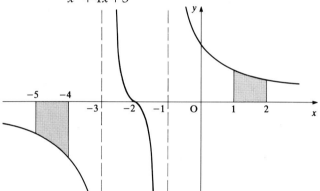

Find the area of each shaded region.

7. The graph of $y = x + \dfrac{4}{x}$ is shown below.

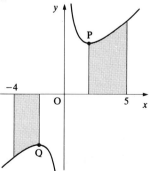

(i) Find the co-ordinates of the minimum point, P, and the maximum point, Q.

(ii) Find the area of each shaded region.

Exercise 5C continued

8. (i) Find $\int_0^X xe^{-x^2}dx$ in terms of X.

   (ii) Evaluate $\int_0^X xe^{-x^2}dx$ for $X = 1,2,3$ and 4. (Give your answers to 4 significant figures.)

   (iii) As X gets bigger (i.e. as $X \to \infty$), towards what value does $\int_0^X xe^{-x^2}dx$ tend?

[MEI]

9. (i) Sketch the curve with equation $y = \dfrac{e^x}{e^x + 1}$ for values of x between 0 and 2.

   (ii) Find the area of the region enclosed by this curve, the axes and the line $x = 2$.

   (iii) Find the value of $\int_1^e \dfrac{2t}{t^2 + 1}dt$.

   (iv) Compare your answers to parts (ii) and (iii). Explain this result.

Investigations

Compound interest

You win £100 000 in a prize draw and are offered two investment options.

A   You are paid 100% interest at the end of 10 years, or

B   You are paid 10% compound interest year by year for 10 years.

Under which scheme are you better off?

Clearly in scheme **A**, the ratio $R = \dfrac{\text{Final Money}}{\text{Original Money}}$ is $\dfrac{£200\,000}{£100\,000} = 2.$

What is the value of the ratio R in scheme **B**?

Suppose that you asked for the interest to be paid in 20 half-yearly instalments of 5% each. What would be the value of R in this case?

Continue this process, investigating what happens to the ratio R when the interest is paid at increasingly frequent intervals.

Is there a limit to R as the time interval between interest payments tends to zero?

A series for e^x

The exponential function can be written as the infinite series

$$e^x = a_0 + a_1 x + a_2 x^2 + a_3 x^3 + a_4 x^4 + \ldots$$

where $a_0, a_1, a_2, \ldots$ are numbers.

You can find the value of a_0 by substituting the value 0 for x. Since $e^0 = 1$, it follows that

$$1 = a_0 + 0 + 0 + 0 + \ldots$$

and so $a_0 = 1$. You can now write

$$e^x = 1 + a_1 x + a_2 x^2 + a_3 x^3 + a_4 x^4 + \ldots$$

Now differentiate both sides

$$e^x = a_1 + 2a_2 x + 3a_3 x^2 + 4a_4 x^3 + \ldots$$

and substitute $x = 0$ again:

$$1 = a_1 + 0 + 0 + 0 + \ldots$$

and so $a_1 = 1$ also.

Now differentiate a second time, and again substitute $x = 0$. This time you find a_2. Continue this procedure until you can see the pattern in the value of $a_0, a_1, a_2, a_3, \ldots$

When you have the series for e^x, substitute $x = 1$. The left hand side is e^1 or e, and so by adding the terms on the right hand side you obtain the value of e. You will find that the terms become small quite quickly, so you will not need to use very many to obtain the value of e correct to several decimal places.

If you are also studying Statistics you will meet this series expansion of e^x in connection with the Poisson distribution.

KEY POINTS

- $\int \dfrac{1}{x} \, dx = \log_e |x| + c$

- $\log_e x$ is called the natural logarithm of x and denoted by $\ln x$.

- $\dfrac{d}{dx}(\ln x) = \dfrac{1}{x}$

- $\int \dfrac{f'(x)}{f(x)} \, dx = \ln|f(x)| + c.$

- $e = 2.7182818284 \ldots$ is the base of natural logarithms.

- e^x and $\ln x$ are inverse functions: $e^{\ln x} = x$; $\ln(e^x) = x$.

- $\dfrac{d}{dx} e^x = e^x$; $\int e^x \, dx = e^x + c$

- $e^x = 1 + x + \dfrac{x^2}{2!} + \dfrac{x^3}{3!} + \dfrac{x^4}{4!} + \ldots$ (This is of importance in Statistics 2.)

6

Numerical solutions of equations

It is the true nature of mankind to learn from his mistakes.

Fred Hoyle

Which of the following equations can be solved algebraically, and which cannot? For each equation find a solution, accurate or approximate.

- $x^2 - 4x + 3 = 0$
- $x^2 + 10x + 8 = 0$
- $x^5 - 5x + 3 = 0$
- $x^3 - x = 0$
- $e^x = 4x$

You probably realised that the equations $x^5 - 5x + 3 = 0$ and $e^x = 4x$ cannot be solved algebraically. You may have decided to draw their graphs, either manually or using a graphics calculator or computer package (see figure 6.1).

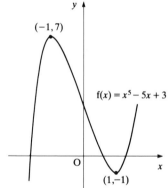

Figure 6.1

The graphs show you that

- $x^5 - 5x + 3 = 0$ has three roots, lying in the intervals $[-2, -1]$, $[0, 1]$ and $[1, 2]$.

- $e^x = 4x$ has two roots, lying in the intervals $[0, 1]$ and $[2, 3]$.

The problem now is how to find the roots to any required degree of accuracy, and as efficiently as possible.

In many real problems, equations are obtained for which solutions using algebraic or analytical methods are not possible, but for which you nonetheless want to know the answers. In this chapter you will be introduced to numerical methods for solving such equations. In applying these methods, keep the following points in mind.

- Only use numerical methods when algebraic ones are not available. If you can solve an equation algebraically (e.g. a quadratic equation), that is the right method to use.

- Before starting to use a calculator or computer program, always start by drawing a sketch graph of the function whose equation you are trying to solve. This will show you how many roots the equation has and their approximate positions. It will also warn you of possible difficulties with particular methods.

- Always give a statement about the accuracy of an answer (e.g. to 5 decimal places, or ± 0.000 005). An answer obtained by a numerical method is worthless without this; the fact that at some point in the procedure your calculator display reads, say, 1.6764705882 does not mean that all these figures are valid.

- Your statement about the accuracy must be obtained from within the numerical method itself. Usually you find a sequence of estimates of ever increasing accuracy.

- Remember that the most suitable method for one equation may not be that for another.

NOTE *An interval written as [a,b] means the interval between a and b, including a and b. This notation is used in this chapter. If a and b are not included, the interval is written]a,b[, or in some books (a,b).*

Interval estimation – change of sign methods

Assume that you are looking for the roots of the equation $f(x) = 0$. This means that you want the values of x for which the graph of $y = f(x)$ crosses the x axis. As the curve crosses the x axis, $f(x)$ changes sign, so provided that $f(x)$ is a continuous function (its graph has no asymptotes or other breaks in it), once you have located an interval in which $f(x)$ changes sign, you know that that interval must contain a root. In both of the graphs in figure 6.2, there is a root lying between a and b.

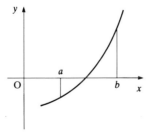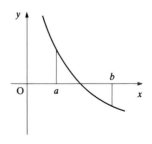

Figure 6.2

You have seen that $x^5 - 5x + 3$ has roots in the intervals $[-2, -1]$, $[0, 1]$ and $[1, 2]$. There are several ways of homing in on such roots systematically. Three of these are now described, using the search for the root in the interval [0, 1] as an example.

Decimal search

In this method you first take increments in x of size 0.1 within the interval [0, 1], working out the value of the function $x^5 - 5x + 3$ for each one. You do this until you find a change of sign.

x	0	0.1	0.2	0.3	0.4	0.5	0.6	0.7
$f(x)$	3.00	2.50	2.00	1.50	1.01	0.53	0.08	-0.33

There is a sign change, and therefore a root, in the interval [0.6, 0.7]. Having narrowed down the interval, you can now continue with increments of 0.01 within the interval [0.6, 0.7].

x	0.60	0.61	0.62
$f(x)$	0.08	0.03	-0.01

This shows that the root lies in the interval [0.61, 0.62].

This process can be continued by considering $x = 0.611$, $x = 0.612$, ... to obtain the root to any required number of decimal places.

For Discussion

How many steps of decimal search would be necessary to find each of the values 0.012, 0.385, and 0.989, using $x = 0$ as a starting point?

When you use this procedure on a computer or calculator you should be aware that the machine is working in base 2, and that the conversion of many simple numbers from base 10 to base 2 introduces small rounding errors. This can lead to simple roots such as 2.7 being missed and only being found as 2.699999.

Interval bisection

This method is similar to the decimal search, but instead of dividing each interval into 10 parts and looking for a sign change, in this case the interval is divided into two parts – it is bisected.

Looking as before for the root in the interval [0, 1], you start by taking the midpoint of the interval, 0.5.

$f(0.5) = 0.53$, so $f(0.5) > 0$. Since $f(1) < 0$, the root is in [0.5, 1].

Now take the midpoint of this second interval, 0.75.

f(0.75) = −0.51, so f(0.75) < 0. Since f(0.5) > 0, the root is in [0.5, 0.75].

The midpoint of this further reduced interval is 0.625:

f(0.625) = −0.03, so the root is in the interval [0.5, 0.625].

The method continues in this manner until any required degree of accuracy is obtained. However, the interval bisection method is quite slow to converge to the root.

Activity

Investigate how many steps you need in this method to achieve accuracy of (i) 1, (ii) 2, (iii) 3, (iv) n decimal places, having started with an interval of length 1.

Linear interpolation

A refinement of this type of method arises when you use not only the signs of the function at the end points of the interval, but its values there as well, to help you to define a reduced interval. As before, an interval (usually of unit length) containing the root is first located. The part of the curve in this interval is then approximated by the chord joining its end points, and the x co-ordinate of the point where the chord crosses the x-axis is calculated.

Looking again at the function $f(x) = x^5 - 5x + 3$, you can see that $f(0) = 3$ and $f(1) = -1$. Figure 6.3 shows the chord of the curve between (0,3) and (1,−1). It crosses the x axis at 0.75.

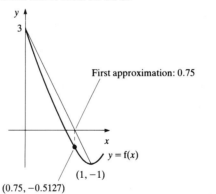
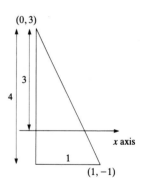

Figure 6.3

The value of $x^5 - 5x + 3$ at $x = 0.75$ is then calculated: f(0.75) = −0.5127.

Since f(0) > 0 and f(0.75) < 0, the root must lie in the interval [0, 0.75].

The procedure is then repeated successively until the required level of accuracy is achieved. The second approximation is shown in figure 6.4.

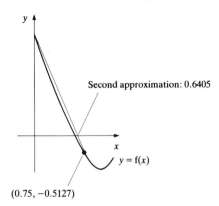

Second approximation: 0.6405

$y = f(x)$

Figure 6.4　　(0.75, −0.5127)

For this example, the method of linear interpolation approaches the root more rapidly than the previous methods, the successive intervals being

[0,1], [0,0.75], [0,0.6405], [0,0.6209], [0,0.6184], [0,0.6181], [0,0.6180].

The sequence of numbers representing the right hand end of the interval appears to be converging, so that you would suspect it to be getting close to the root. (It is not always the right hand end of the interval that does this: in other examples, it may be the left hand end.) The left hand end of the interval is still far from the root, however, so you cannot be sure of your level of accuracy. You now need to see if you can move the left hand end much closer to the suspected root without a change of sign. You might look at $x = 0.6179$, for example. It turns out that f(0.6179) is positive, so the interval for the root is now closed down to [0.6179, 0.6180].

It is difficult to predict the number of steps of linear interpolation which would be needed to reach any required level of accuracy. The rate at which the root is found (the *rate of convergence*) is very variable for this method, as shown in figure 6.5.

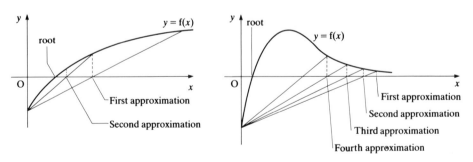

Rapid rate of convergence　　　　　　　Slow rate of convergence

Figure 6.5

Activity

Given that the equation f(x) = 0 has a root between $x = a$ and $x = b$, show that linear interpolation would next lead you to investigate

$$x = \frac{bf(a) - af(b)}{f(a) - f(b)}.$$

Comments on change of sign methods

Error (or solution) bounds

Change of sign methods have the great advantage that they automatically provide bounds (the two ends of the interval) within which a root lies, so the maximum possible error in a result is known. Knowing that a root lies in the interval [0.61, 0.62] means that you can take the root as 0.615 with a maximum error of ± 0.005.

Problems with change of sign methods

There are a number of situations which can cause problems for change of sign methods if they are applied blindly, for example by entering the equation into a computer program without prior thought. In all cases you can avoid problems by first drawing a sketch graph, provided that you know what dangers to look out for.

(i)   *The curve touches the x axis*

In this case there is no change of sign, so change of sign methods are doomed to failure (figure 6.6).

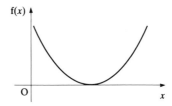

Figure 6.6

(ii)   *There are several roots close together*

Where there are several roots close together, it is easy to miss a pair of them. The equation

$$f(x) = x^3 - 1.9x^2 + 1.11x - 0.189 = 0$$

has roots at 0.3, 0.7 and 0.9. A sketch of the curve of f(x) is shown in figure 6.7.

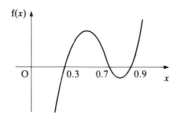

Figure 6.7

In this case f(0) < 0 and f(1) > 0, so you know there is a root between 0 and 1.

A decimal search would show that f(0.3) = 0, so that 0.3 is a root. You would be unlikely to search further in this interval.

Interval bisection gives f(0.5) > 0, so you would search the interval [0, 0.5] and eventually arrive at the root 0.3, unaware of the existence of those at 0.7 and 0.9. Linear interpolation would give you 0.9 only.

(iii) *There is a discontinuity in f(x)*

The curve $y = \dfrac{1}{x - 2.7}$ has a discontinuity at $x = 2.7$, as shown in figure 6.8.

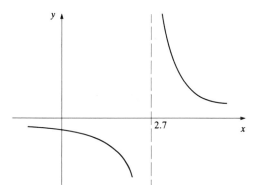

Figure 6.8

The equation $\dfrac{1}{x - 2.7} = 0$ has no root, but all change of sign methods will converge on a false root at $x = 2.7$.

None of these problems will arise if you start by drawing a sketch graph.

Exercise 6A

1. Use (i) decimal search, (ii) interval bisection and (iii) linear interpolation to find the roots of $x^5 - 5x + 3 = 0$ in the intervals $[-2, -1]$ and $[1, 2]$, correct to two decimal places.

   Comment on the ease and efficiency with which the roots are approached by each method.

2. (i)   Use a systematic search for a change of sign, starting with $x = -2$, to locate intervals of unit length containing each of the three roots of $x^3 - 4x^2 - 3x + 8 = 0$.
   (ii) Sketch the graph of $f(x) = x^3 - 4x^2 - 3x + 8$.
   (iii) Use the method of interval bisection to obtain each of the roots correct to two decimal places.

3. The diagram shows a sketch of the graph of $f(x) = e^x - x^3$ without scales.

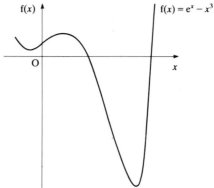

f(x)

$f(x) = e^x - x^3$

O

x

(i) Use a systematic search for a change of sign to locate intervals of unit length containing each of the roots.

(ii) Use linear interpolation to find each of the roots correct to three decimal places.

4. (i) Show that the equation $x^3 + 3x - 5 = 0$ has no turning points.

(ii) Show with the aid of a sketch that the equation can have only one root, and that this root must be positive.

(iii) Find the root, correct to three decimal places.

5. (i) How many roots has the equation
$$e^x - 3x = 0$$

(ii) Find an interval of unit length containing each of the roots.

(iii) Find each root correct to two decimal places.

6. (i) Sketch $y = 2^x$ and $y = x + 2$ on the same axes.

(ii) Use your sketch to deduce the number of roots of the equation
$2^x = x + 2$.

(iii) Find each root, correct to three decimal places if appropriate.

7. Find all the roots of $x^3 - 3x + 1 = 0$, giving your answers correct to two decimal places.

8. For each of the equations below,

(i) Sketch the curve;

(ii) Write down any roots;

(iii) investigate what happens when you use a change-of-sign method with a starting interval of $[-0.3, 0.7]$.

(a) $y = \dfrac{1}{x}$ (b) $y = \dfrac{x}{x^2 + 1}$ (c) $y = \dfrac{x^2}{x^2 + 1}$

Fixed point estimation

In fixed point estimation you find a single value or point as your estimate for the value of x, rather than establishing an interval within which it must lie. This involves an *iterative process*, a method of generating a sequence of numbers by continued repetition of the same procedure. If the numbers obtained in this manner approach some limiting value, then they are said to *converge* to this value.

Investigation

Notice what happens in each of the following cases, and try to find some explanation for it.

(i)   Set your calculator to the radian mode and press the cosine button repeatedly.

(ii) Enter any positive number into your calculator and press the square root button repeatedly. Try this for both large and small numbers.

(iii) Enter any positive number into your calculator and press the sequence $\boxed{+}\ \boxed{1}\ \boxed{=}\ \boxed{\sqrt{}}$ repeatedly. Write down the number which appears each time you press $\boxed{\sqrt{}}$. The sequence generated appears to converge. You may recognise the number to which it appears to converge: it is called the Golden Ratio.

Two methods of fixed point estimation are introduced in this chapter: the first one involves rearranging the equation to be solved into the form $x = g(x)$. The second is called the Newton–Raphson method; it is actually a special case of rearranging the equation, but it is treated as a separate method here.

Rearranging the equation f(x)=0 into the form $x = g(x)$

The first step, with an equation f(x) = 0, is to rearrange it into the form $x = g(x)$. Any value of x for which $x = g(x)$ is clearly a root of the original equation, as shown in figure 6.9.

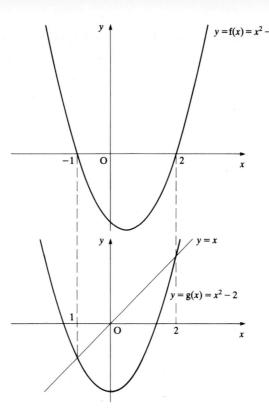

Figure 6.9

The equation $x^5 - 5x + 3 = 0$ which you met earlier can be rewritten in a number of ways. One of these is

$$x = g(x) = \frac{x^5 + 3}{5}.$$

Figure 6.10 shows the graphs of $y = x$ and $y = g(x)$ in this case.

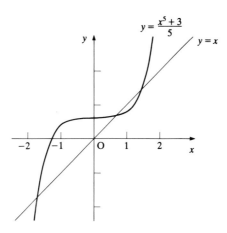

Figure 6.10

This provides the basis for the iterative formula

$$x_{n+1} = \frac{x_n{}^5 + 3}{5}.$$

Taking $x = 1$ as a starting point to find the root in the interval $[0, 1]$, successive approximations are

$x_1 = 1$, $\qquad x_2 = 0.8$, $\qquad x_3 = 0.6655$, $\qquad x_4 = 0.6261$, $\qquad x_5 = 0.6192$,
$x_6 = 0.6182$, $\qquad x_7 = 0.6181$, $\qquad x_8 = 0.6180$, $\qquad x_9 = 0.6180$.

In this case the iteration has converged quite rapidly to the root for which you were looking.

For Discussion

Another way of arranging $x^5 - 5x + 3 = 0$ is $x = \sqrt[5]{(5x - 3)}$. What other possible rearrangements can you find?

The iteration process is easiest to understand if you consider the graph. Rewriting the equation $f(x) = 0$ in the form $x = g(x)$ means that instead of looking for points where the graph of $y = f(x)$ crosses the x axis, you are now finding the points of intersection of the curve $y = g(x)$ and the line $y = x$.

What you do	**What it looks like on the graph**
● Choose a value, x_1, of x	Take a starting point on the x axis
● Find the corresponding value of $g(x_1)$	Move vertically to the curve $y = g(x)$
● Take this value $g(x_1)$ as the new value of x, i.e. $x_2 = g(x_1)$	Move across to the line $y = x$
● Find the value of $g(x_2)$	Move vertically to the curve

and so on.

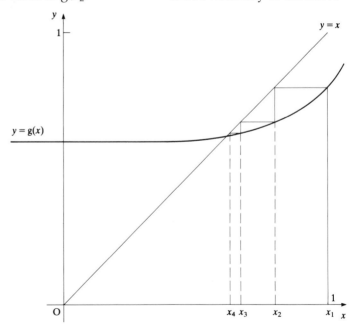

Figure 6.11

The effect of several repeats of this procedure is shown in figure 6.11. The successive steps look like a staircase approaching the root: this type of diagram is called a *staircase diagram*. In other examples, a *cobweb diagram* may be produced, as shown in figure 6.12.

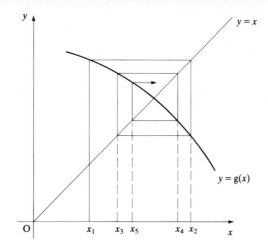

Figure 6.12

Successive approximations to the root are found by using the formula

$$x_{n+1} = g(x_n).$$

This is an example of an *iterative formula*. If the resulting values of x_n approach some limit, a, then $a = g(a)$, and so a is a *fixed point* of the iteration. It is also a root of the original equation $f(x) = 0$.

Notice that in the staircase diagram, the values of x_n approach the root from one side, but in a cobweb diagram they oscillate about the root. From figures 6.11 and 6.12 it is clear that the error (the difference between a and x_n) is decreasing in both diagrams.

Using different arrangements of the equation

So far we have used only one possible arrangement of the equation $x^5 - 5x + 3 = 0$. What happens when you use a different arrangement, for example $x = \sqrt[5]{(5x - 3)}$, which leads to the iterative formula

$$x_{n+1} = \sqrt[5]{(5x_n - 3)}?$$

The resulting sequence of approximations is

$x_1 = 1,$ $x_2 = 1.1486...,$ $x_3 = 1.2236...,$ $x_4 = 1.2554...,$
$x_5 = 1.2679...,$ $x_6 = 1.2727...,$ $x_7 = 1.2745...,$ $x_8 = 1.2752...,$
$x_9 = 1.2755...,$ $x_{10} = 1.2756...,$ $x_{11} = 1.2756...,$ $x_{12} = 1.2756....$

NOTE

In these calculations the full calculator values of x_n were used, but only the first four decimal places have been written down.

The process has clearly converged, but in this case not to the root for which you were looking: you have identified the root in the interval [1, 2]. If instead you had taken $x_1 = 0$ as your starting point and applied the second formula, you would have obtained a sequence converging to the value -1.6180, the root in the interval $[-2, -1]$.

The choice of g(x)

A particular rearrangement of the equation $f(x) = 0$ into the form $x = g(x)$ will allow convergence to a root a of the equation, provided that $-1 < g'(a) < 1$ for values of x close to the root.

Look again at the two rearrangements of $x^5 - 5x + 3 = 0$ which were suggested. When you look at the graph of

$$y = g(x) = \sqrt[5]{(5x - 3)},$$

you can see that its gradient near A, the root you were seeking, is greater than 1 (figure 6.13). This makes

$$x_{n+1} = \sqrt[5]{(5x_n - 3)}$$

an unsuitable iterative formula for finding the root in the interval [0,1], as you saw earlier.

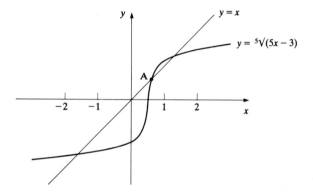

Figure 6.13

When an equation has two or more roots, a single rearrangement will not usually find all of them. This is demonstrated in figure 6.14.

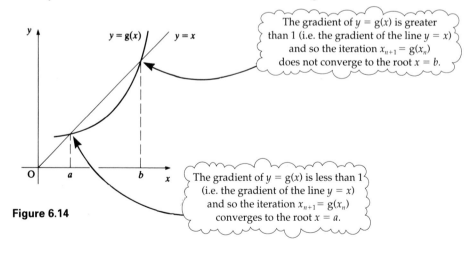

The gradient of $y = g(x)$ is greater than 1 (i.e. the gradient of the line $y = x$) and so the iteration $x_{n+1} = g(x_n)$ does not converge to the root $x = b$.

The gradient of $y = g(x)$ is less than 1 (i.e. the gradient of the line $y = x$) and so the iteration $x_{n+1} = g(x_n)$ converges to the root $x = a$.

Figure 6.14

Activity

Try using the iterative formula $x_{n+1} = \dfrac{x_n^5 + 3}{5}$ to find the roots in the intervals $[-2, -1]$ and $[1, 2]$. In both cases use each end point of the interval as a starting point. What happens?

Explain what you find by referring to a sketch of the curve $y = \dfrac{x^5 + 3}{5}$.

Accuracy of method of rearranging equation

Iterative procedures give you a sequence of point estimates. A staircase diagram, for example, might give the following:

$$1, 0.8, 0.6655, 0.6261, 0.6192$$

What can you say at this stage?

Looking at the pattern of convergence it seems as though the root lies between 0.61 and 0.62, but you cannot be absolutely certain from the available evidence. To be certain you must look for a change of sign.

$$f(0.61) = +\,0.034\ldots \qquad f(0.62) = -\,0.0083\ldots$$

Now you can be quite certain that your judgement is correct. You have established bounds for the root.

Estimates from a cobweb diagram oscillate above and below the root and so naturally provide you with bounds.

When does this method fail?

It is always possible to rearrange an equation $f(x) = 0$ into the form $x = g(x)$, but this only leads to a successful iteration if

(i)   successive iterations converge;

(ii) they converge to the root for which you are looking.

Exercise 6B

1. (i)   Show that the equation $x^3 - x - 2 = 0$ has a root between 1 and 2.
   (ii) The equation is rearranged into the form $x = g(x)$, where $g(x) = \sqrt[3]{(x + 2)}$. Show that $-1 < g'(x) < 1$ for values of x in the interval $[1, 2]$.

   (iii) Use the iterative formula suggested by this rearrangement to find the value of the root to three decimal places.

2. (i)   Show that the equation $e^{-x} - x + 2 = 0$ has a root in the interval $[2, 3]$.
   (ii) The equation is rearranged into the form $x = g(x)$ where $g(x) = e^{-x} + 2$. Show that $-1 < g'(x) < 1$ for values of x in the interval $[2, 3]$.

Exercise 6B continued

    (iii) Use the iterative formula suggested by this rearrangement to find the value of the root to three decimal places.

3. (i) By considering $f'(x)$, where $f(x) = x^3 + x - 3$, show that there is exactly one real root of the equation $x^3 + x - 3 = 0$.

    (ii) Show that the root lies in the interval $[1, 2]$.

    (iii) Rearrange the equation into the form $x = g(x)$ where $-1 < g'(x) < 1$ for values of x close to the root.

    (iv) Hence find the root correct to four decimal places.

4. (i) Show that the equation $e^x + x - 6$ has a root in the interval $[1, 2]$.

    (ii) Show that this equation may be written in the form $x = \ln(6 - x)$.

    (iii) Hence find the root correct to three decimal places.

5. (i) Sketch the curves $y = e^x$ and $y = x^2 + 2$ on the same graph.

    (ii) Use your sketch to explain why the equation $e^x - x^2 - 2 = 0$ has only one root.

    (iii) Rearrange this equation in the form $x = g(x)$.

    (iv) Find the root correct to three decimal places

6. (i) Show that $x^2 = \ln(x + 1)$ for $x = 0$ and for one other value of x.

    (ii) Use the method of fixed point estimation to find the second value to three decimal places.

7. (i) Sketch the graphs of $y = x$ and $y = \cos x$ on the same axes, for
$$0 \le x \le \tfrac{\pi}{2}.$$

    (ii) Find the solution of the equation $x = \cos x$ to five decimal places.

Activity

(i)   Show that the equation $\ln x - \sin x = 0$ has only one root.

(ii) Rearrange the equation in the form $x = g(x)$.

(iii) Explain your results when you try to find the root using the iteration
$$x_{n+1} = g(x_n).$$

The Newton–Raphson method

This is another fixed point estimation method, and as for the previous method it is necessary to use an estimate of the root as a starting point.

You start with an estimate, x_1, for a root of $f(x) = 0$. You then draw the tangent to the curve $y = f(x)$ at the point $(x_1, f(x_1))$. The point at which the tangent cuts the x axis then gives the next approximation for the root, and the process is repeated (figure 6.15).

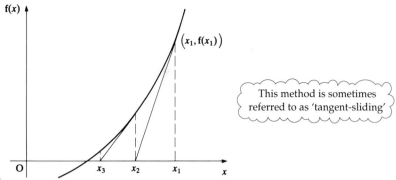

Figure 6.15

The gradient of the tangent at $(x_1, f(x_1))$ is $f'(x_1)$. Since the equation of a straight line can be written

$$y - y_1 = m(x - x_1),$$

the equation of the tangent is

$$y - f(x_1) = f'(x_1)[x - x_1].$$

This tangent cuts the x axis at $(x_2, 0)$, so

$$0 - f(x_1) = f'(x_1)[x_2 - x_1]$$

Rearranging this,

$$x_2 = x_1 - \frac{f(x_1)}{f'(x_1)}.$$

This gives rise to the Newton–Raphson iterative formula

$$x_{n+1} = x_n - \frac{f(x_n)}{f'(x_n)}.$$

Returning to the equation $x^5 - 5x + 3 = 0$, which has a root in the interval $[0, 1]$, you can write

$$f(x) = x^5 - 5x + 3 \qquad \text{and} \qquad f'(x) = 5x^4 - 5$$

The iterative formula is therefore

$$x_{n+1} = x_n - \frac{f(x_n)}{f'(x_n)}$$

$$= x_n - \frac{x_n^5 - 5x_n + 3}{5x_n^4 - 5}.$$

Starting with $x_1 = 0$ gives

$x_2 = 0.6, \qquad x_3 = 0.6178676\ldots, \qquad x_4 = 0.6180339\ldots, \qquad \ldots$

which is a faster rate of convergence than any of the earlier methods gave.

Investigation

What happens when you try $x_1 = 1$ as a starting point in the iteration

$$x_{n+1} = -\frac{x_n^5 - 5x_n + 3}{5x_n^4 - 5} \; ?$$

Illustrate this on a graph.

In this example the Newton–Raphson method gives an extremely rapid rate of convergence. This is the case for most examples, even when the first approximation is not particularly good. A discussion of the rate of convergence of this method is beyond the scope of this text, but for manual calculations it is almost always the most efficient method.

Problems with the Newton–Raphson method

Most problems that arise with the Newton–Raphson method fall into one or other of the following two categories.

(i) *Poor choice of starting value*

If your initial value is close enough to a root, the method will nearly always give convergence to it. However if the initial value is not close to the root, or is near a turning point of $y = f(x)$, the iteration may diverge, or converge to another root.

When the first approximation is close to a turning point, $f'(x_1)$ will be very small. In most cases this will mean that x_2 is not very close to the root (see figure 6.16).

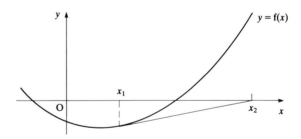

Figure 6.16

In this case you may find that after two or three steps the values you compute are converging rapidly, but they may be converging to a root other than the one which you are trying to locate.

(ii) *The function is discontinuous*

As with all numerical methods for solving equations, this method can break down when the equation is that of a discontinuous function.

Exercise 6C

1. (i) Sketch the curve $f(x) = \dfrac{x^3}{3} - x + 2$.

  (ii) Using the Newton–Raphson method find the root of the equation $f(x) = 0$, starting with $x_1 = -5$, correct to 3 decimal places.

  (iii) What happens if the starting value is taken to be $x_1 = 0.5$?

2. (i) Show that the equation $x^4 - 7x^3 + 1 = 0$ has a root in the interval $[0, 1]$.

  (ii) Use the Newton – Raphson method to find this root correct to 2 decimal places, starting with $x_1 = 1$.

  (iii) Explain why $x_1 = 0$ is not a suitable starting point.

3. (i) Show that the equation $e^x - 3x^2 = 0$ has three roots in the interval $[-1, 4]$.

  (ii) Use the Newton–Raphson method to find each of the roots correct to 2 decimal places. In each case state the starting value which gave covergence to the particular root.

  (iii) A starting value $x_1 = 0$, gives $x_2 = -1$. Explain this result.

4. (i) Show that the equation $x^2 - 3x \ln x = 0$ has two roots in the interval $[1, 5]$.

  (ii) Use the Newton–Raphson method to find each root correct to 2 decimal places.

5. Using the Newton–Raphson method or otherwise find, correct to 3 decimal places, the value of x for which $x = e^{-x}$.

                                             [MEI]

6. (i) Show that the equation $x^3 - x^2 - 2x + 1 = 0$ has three roots in the interval $[-2, 2]$.

  (ii) Use the Newton–Raphson method to find each of the three roots correct to 4 decimal places.

  (iii) Investigate whether the root found is always that nearest the starting point.

7. (i) Show that the equation $x^3 - 3x^2 + 1 = 0$ has exactly three roots.

  (ii) Use the Newton–Raphson method to find each of the two smaller roots correct to three decimal places.

  (iii) Find the largest root to the limit of the accuracy of your calculator, using $x_1 = 20$ as starting point.

  (iv) Investigate how many digits the method has fixed after each iteration, and comment on your findings.

8. (i) Sketch the curves $y = e^x$ and $y = \dfrac{4}{x}$. At how many points do they intersect?

Exercise 6C continued

    (ii) Sketch the graph of the function $\dfrac{4}{x} - e^x$ for all values of x.

    (iii) Use the Newton–Raphson method to find the value of x where the curve $y = \dfrac{4}{x} - e^x$ crosses the x axis, correct to 3 decimal places, taking $x_1 = 2$.

    (iv) Explain what happens if you use a starting value of $x_1 = -3$.

The following investigations illustrate cases where problems arise when using the Newton–Raphson method. In each case finish the investigation by suggesting how the roots can be found as easily as possible.

Investigations

1. The function $f(x) = \ln (x + 2) - x$ is not defined for $x \le -2$. The line $x = -2$ is an asymptote, as shown in the diagram.

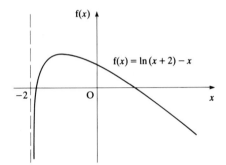

A systematic search for a sign change reveals that there are roots in the intervals $[-2, -1]$ and $[1, 2]$.

Using the Newton–Raphson method with $x_1 = -1$, try to find the smaller root. Describe what happens.

Now try $x_1 = -1.5$. What happens now?

2. When using the sign change principle to locate the roots of $f(x) = 0$, where

$$f(x) = 9 - \frac{1}{x^2 - 4x + 4.1}$$

the following results are obtained.

x	0	1	2	3	4
$f(x)$	8.76	8.09	-1	8.09	8.76

This shows that there are roots in each of the intervals $[1, 2]$ and $[2, 3]$.

Investigate what happens when the Newton–Raphson method is used to find the smaller root, and complete the table below.

Starting point	1	1.2	1.4	1.6	1.8	2
Result						

KEY POINTS

Interval estimation

- When $f(x)$ is a continuous function, if $f(a)$ and $f(b)$ have opposite signs, there will be at least one root of $f(x) = 0$ in the interval $[a, b]$.

- When an interval $[a, b]$ containing a root has been found, this interval may be reduced systematically by one of the following methods:

  - a decimal search within the interval,

  - interval bisection,

  - linear interpolation.

- Solution bounds are provided automatically by these methods.

Fixed point estimation

- Fixed point estimation may be used to solve an equation $f(x) = 0$ by either of the following methods:

  - rearranging the equation $f(x) = 0$ into the form $x = g(x)$ where $-1 < g'(x) < 1$ near the root, and using the iteration

  $$x_{n+1} = g(x_n).$$

  - the Newton–Raphson method using the iteration

  $$x_{n+1} = x_n - \frac{f(x_n)}{f'(x_n)}.$$

- Solution bounds are usually confirmed by demonstrating a change of sign of $f(x)$ between them.

Answers

Exercise 1A

1. (a) 3^4 (b) 3^0 (c) 3^3 (d) 3^{-3} (e) $3^{\frac{1}{2}}$ (f) 3^1 2. (a) 4^2 (b) $4^{\frac{1}{2}}$ (c) 4^{-1} (d) $4^{-\frac{1}{2}}$ (e) $4^{\frac{1}{2}}$ (f) $4^{-\frac{1}{2}}$
3. (a) 10^3 (b) 10^{-4} (c) 10^{-3} (d) $10^{\frac{1}{2}}$ (e) $10^{-\frac{1}{2}}$ (f)10^{-6} 4. (a) $\frac{1}{32}$ (b) 3 (c) $\frac{1}{3}$ (d) 5 (e) $\frac{1}{5}$ (f) 1
5. (a) 9 (b) 10 (c) $\frac{1}{5}$ (d) 1 (e) $5^5 = 3125$ (f) 27   6. (a) 49 (b) 3 (c) 3 (d) $3^6 = 729$ (e) 8 (f) 4
7. (a) 4 (b) 0 (c) 0 (d) $1\frac{15}{16}$ (e) 0 (f) $2\frac{2}{3}$

Exercise 1B

1. (a) $\sqrt{7}$ (b) $\sqrt{3}$ (c) $\sqrt{2}+1$ (d) 1 (e) $2(\sqrt{3}-\sqrt{2})$
2. (a) $8+3\sqrt{2}$ (b) $5\sqrt{2}$ (c) $19+16\sqrt{5}$ (d) $8\sqrt{2}+2\sqrt{5}$ (e) $6\sqrt{2}$
3. (a) $10+\sqrt{2}$ (b) $7-4\sqrt{3}$ (c) $10-2\sqrt{5}$ (d) -4 (e) $1+\frac{1}{2}\sqrt{3}$ (f) $3\sqrt{7}-9$ (g) 14 (h) $\sqrt{30}$ (i) 2
4. (a) $1+\frac{1}{2}\sqrt{2}$ (b) $\frac{1}{2}\left(3\sqrt{3}-5\right)$ (c) $1-\frac{2}{5}\sqrt{5}$ (d) $-\frac{1}{4}\left(1+\sqrt{5}\right)$ (e) $-3-\sqrt{3}$ (f) $\frac{1}{3}\left(\sqrt{7}+1\right)$ (g) $\frac{1}{7}\left(9+4\sqrt{2}\right)$ (h) $11+2\sqrt{3}$

Exercise 1C

1. (a) 4 (b) -4 (c) $\frac{1}{2}$ (d) 0 (e) 4 (f) -4 (g) $\frac{3}{2}$ (h) $\frac{1}{4}$ (i) $\frac{1}{2}$ (j) -3
2. (a) $\log 10$ (b) $\log 2$ (c) $\log 36$ (d) $\log \frac{1}{7}$ (e) $\log 3$ (f) $\log 4$ (g) $\log 4$ (h) $\log \frac{1}{3}$ (i) $\log \frac{1}{2}$ (j) $\log 12$
3. (a) $n = 19.93$ (b) $n = -9.97$ (c) $n = 9.01$ (d) $n -48.32$ (e) $n= 1375$ 4. 11

Exercise 1D

The techniques in this exercise involve drawing a line of best fit by eye. Consequently your answers may reasonably vary a little from those given.

1. (i) If the relationship is of the form $R = kT^n$, the graph of log R against log T will be a straight line.
   (ii) log R 5.46   5.58   5.72   6.09   6.55
        log T 0.28   0.43   0.65   1.20   1.90

   (iii) $k = 1.8 \times 10^5$, $n = 0.690$
   (iv) 0.7 days

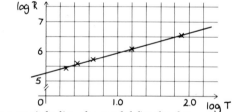

2. (ii) Plotting log A against t will test the model: if it is a straight line the model fits the data.
   (iii)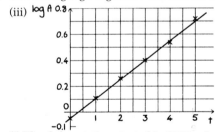

   (iii contd.) $b = 1.4$, $k = 0.89$
   (iv) (a) $t = 2.4$ days
        (b) $A = 3.0$ cm^2
   (v) Exponential growth

3. (i) The deforestation started in 1910, when there were 3200 trees.
   (ii) The constant k is the original number of trees.
   The value of a will be less than 1 because the number of trees is decreasing.
   (iv)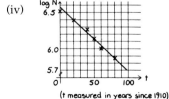
        (t measured in years since 1910)

   (iv contd.) $k = 3.7 \times 10^6$, $a = 0.98$

   (v) In the year 2009.

4. (i)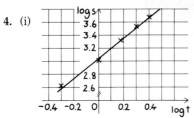

(i contd.) The model $s = kt^n$ can be written as $\log s = \log k + n \log t$, so if this model is appropriate the graph of $\log s$ against $\log t$ will be a straight line. Since the points do lie in a straight line, the model is appropriate.

(ii) $k = 1100$, $n = 1.6$ (iii) $s = 2500$ m

(iv) The train would not continue to accelerate like this throughout its journey. After 10 minutes it would probably be travelling at constant speed, or possibly even slowing down.

5. Taking logs of both sides, $\log y = \log A + B \log x$.
Plotting $\log y$ against $\log x$ gives a straight line of gradient B and intercept $\log A$.
This gives $A = 1.5$, $B = 0.78$.
The value of y that is wrong is 6.21. If x is 5.07, y should be 5.32 according to the equation.

6. $k = 1.3$, $n = 0.50$

7. $\log y = B \log x + \log A$.

He should plot $\log y$ against $\log x$. If this gives a straight line, there is a relationship of the form $y = Ax^B$. If there is no such relationship, the points will not be in a straight line.
The value of $\log A$ is given by the intercept on the $\log y$ axis. The value of B is the gradient of the line.

$\log x$	0.60	0.85	1.00	1.11	1.30
$\log y$	0.48	0.60	0.68	0.73	0.83

From the graph, $A = 1.5$, $B = 0.5$.
The formula is therefore $y = 1.5x^{0.5}$.

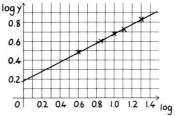

Exercise 2A

1. (a) 19 22 25 28. Arithmetic with 1st term $a = 7$ and common difference $d = 3$.
(b) 4 3 2 1. Arithmetic: $a = 8$, $d = -1$.
(c) 810 2430 7290 21870. Geometric with 1st term $a = 10$ and common ratio $r = 3$.
(d) 4 2 1 $\frac{1}{2}$. Geometric: $a = 64$, $r = \frac{1}{2}$ (e) 2 2 2 5. Periodic with period $p = 4$.
(f) 4 2 1 2. Periodic: $p = 8$ (g) −32 64 −128 256. Geometric, oscillating, non-periodic: $a = 1$, $r = -2$.
(h) 4 6 8 2. Periodic: $p = 4$ (i) 3.3 3.1 2.9 2.7. Arithmetic: $a = 4.1$, $d = -0.2$.
(j) 1.4641 1.61051 1.771561 1.9487171. Geometric: $a = 1$, $r = 1.1$

2. (a) 3 5 7 9 (b) 6 12 24 48 (c) 4 8 14 24 (d) 1 $\frac{1}{2}$ $\frac{1}{3}$ $\frac{1}{4}$ (e) 12 15 18 21
(f) −5 5 −5 5 (g) 72 36 18 9 (h) 4 6 4 6 (i) 4 6 4 6 (j) 1 4 9 16

3. (a) 58 (b) 90 (c) 25 (d) 40 (e) 8

4. (a) $\sum\limits_{1}^{10} k$ (b) $\sum\limits_{1}^{10}(20 + k)$ (c) $\sum\limits_{1}^{10}(200 + 10k)$ (d) $\sum\limits_{1}^{10}(200 + 11k)$ (e) $\sum\limits_{1}^{10}(200 + 10k)$

5. (a) 15 (b) 401 (c) −20

6. (i) 0 1 1 2 3. The sequence of differences, after the first term, is the same as the sequence itself.
(ii) 13 21 34 (iii) 1.0 2.0 1.5 1.667 1.6 1.625 1.615 1.619. The ratios are converging.

7. (i) 2 6 2 6 2 6 (ii) Oscillating periodic sequence, $p = 2$

(iii) (a) 2 8 −4 20 −28 68. Diverging oscillating sequence.

(b) $3\frac{1}{2}$ $4\frac{1}{4}$ $3\frac{7}{8}$ $4\frac{1}{16}$ $3\frac{31}{32}$ $4\frac{1}{64}$. Oscillating sequence, converging towards 4.

8. (i) 4.000 2.667 3.467 2.895 3.340 2.976 3.284 3.017.

(ii) The numbers appear to converge towards a value between 3.0 and 3.2.

9. (i) $\frac{1}{3} + \frac{1}{4} = \frac{7}{12} > \frac{1}{2}$; $\frac{1}{5} + \frac{1}{6} + \frac{1}{7} + \frac{1}{8} = \frac{1066}{1080} > \frac{1}{2}$. This is true for each bracket formed, so $S > 1 + \frac{1}{2} + \frac{1}{2} + \frac{1}{2} + \cdots$

(ii) $1 + \frac{1}{2} + \frac{1}{2} + \frac{1}{2} + \cdots$ is clearly infinite, and S is greater than this.

Exercise 2B

1. (a) Yes: $d = 2$, $a_7 = 39$ (b) No (c) No (d) Yes: $d = 4$, $a_7 = 27$ (e) Yes: $d = -2$, $a_7 = -4$
2. (i) 10 (ii) 37 **3.** (i) 4 (ii) 34 **4.** (i) 5 (ii) 850 **5.** (i) 16 18 20 (ii) 324
6. (i) 15 (ii) 1170 **7.** (i) First term 4, common difference 6 (ii) 12
8. (i) 3 (ii) 165 **9.** (i) 23p (ii) £1.44
10. (i) 5000 (ii) 5100 (iii) 10 100 (iv) The 1st sum, 5000, and the 2nd sum, 5100, add up to the third sum, 10 100. This is because the sum of the odd numbers plus the sum of the even numbers between 50 and 150 is the same as the sum of all the numbers between 50 and 150.
11. (i) 22000 (ii) The 17th term is the first negative term.
12. (i) £16500 (ii) 8 **13.** (i) 49 (ii) 254.8km **14.** (i) 16 (ii) 2.5cm

Exercise 2C

1. (a) Yes: $r = 2$, $a_7 = 320$ (b) No (c) Yes: $r = -1$, $a_7 = 1$ (d) Yes: $r = 1$, $a_7 = 5$ (e) No (f) Yes: $r = \frac{1}{2}$, $a_7 = \frac{3}{32}$ (g) No
2. (i) 384 (ii) 765 **3.** (i) 4 (ii) 81920 **4.** (i) 9 (ii) 10th term **5.** (i) 9 (ii) 4088
6. (i) 6 (ii) 267 (to 3 sig. figs.) **7.** (i) 2 (ii) 3 (iii) 3069 **8.** (i) $\frac{1}{2}$ (ii) 8 **9.** (i) $\frac{1}{10}$ (ii) $\frac{7}{9}$
10. (i) 0.9 (ii) 45th (iii) 1000 (iv) 44 **11.** (i) 0.2 (ii) 1
12. (i) 20 10 5 2.5 1.25 (ii) 0 10 15 17.5 18.75 (iii) First series geometric, common ratio $\frac{1}{2}$. Second sequence not geometric as there is no common ratio.
13. (i) 68th swing is the first less than 1°. (ii) 241° (to nearest degree)
14. (i) Height after nth impact $= 10 \times \left(\frac{2}{3}\right)^n$ (ii) 59.0m (to 3 sig. figs.) **15.** 160m (ii) £933 000
16. (i) Common ratio $r = 1.08$ (ii) £29 985 (iv) £1496 (nearest £)

Exercise 3A

1. (a) One-to-one, yes, equal (b) Many-to-one, yes, not equal (c) Many-to-many, no, equal (d) One-to-many, no, equal (e) Many-to-many, no, not equal (f) One-to-one, yes, not equal (g) Many-to-many, no, equal (h) Many-to-one, yes, not equal
2. (a) (i) Examples: one $\to$ 3, word $\to$ 4 (ii) Many-to-one (iii) Domain: words, co-domain: $\mathbb{Z}$
   (b) (i) Examples: $1 \to 4$, $2.1 \to 8.4$ (ii) One-to-one (iii) Domain: $\mathbb{R}^+$, co-domain: $\mathbb{R}^+$
   (c) (i) Examples: $1 \to 1$, $6 \to 4$ (ii) Many-to-one (iii) Domain: $\mathbb{Z}^+$, co-domain: $\mathbb{Z}^+$
   (d) (i) Examples: $1 \to -3$, $-4 \to -13$ (ii) One-to-one (iii) Domain: $\mathbb{R}$, co-domain: $\mathbb{R}$
   (e) (i) Examples: $4 \to 2$, $9 \to 3$ (ii) One-to-one (iii) Domain: $x \geq 0$, co-domain: $x \geq 0$
   (f) (i) Examples: $36\pi \to 3$, $\frac{9}{2}\pi \to 1.5$ (ii) One-to-one (iii) Domain: $\mathbb{R}^+$, co-domain: $\mathbb{R}^+$
   (g) (i) Examples: $12\pi \to 3$, $12\pi \to 12$ (ii) Many-to-many (iii) Domain: $\mathbb{R}^+$, co-domain: $\mathbb{R}^+$
   (h) (i) Examples: $1 \to \frac{3}{2}\sqrt{3}$, $4 \to 24\sqrt{3}$ (ii) One-to-one (iii) Domain: $\mathbb{R}^+$, co-domain: $\mathbb{R}^+$
   (i) (i) Examples: $4 \to 16$, $-0.7 \to 0.49$ (ii) Many-to-one (iii) Domain: $\mathbb{R}$, co-domain: $x \geq 0$
3. (a) (i) -5 (ii) 9 (iii) -11 (b) (i) 3 (ii) 5 (iii) 10 (c) (i) 32 (ii) 82.4 (iii) 14 (iv) -40
4. (a) $f(x) \leq 2$ (b) $0 \leq f(\theta) \leq 1$ (c) $y \in \{2,3,6,11,18\}$ (d) $y \in \mathbb{R}^+$ (e) $\mathbb{R}$ (f) $\{\frac{1}{2}, 1, 2, 4\}$ (g) $0 \leq y \leq 1$ (h) $f(\theta) \geq 1$ or $f(\theta) \leq -1$ (i) $0 < f(x) \leq 1$ (j) $f(x) \geq 3$
5. For f, every value of x (including $x = 3$) gives a unique output, whereas g(2) can equal either 4 or 6.

Exercise 3B

1. (a) Translation $\begin{pmatrix} 0 \\ -2 \end{pmatrix}$; $x = 0$ (b) Translation $\begin{pmatrix} -4 \\ 0 \end{pmatrix}$; $x = -4$
   (c) Stretch parallel to y axis of s.f. 4, or stretch parallel to x axis of s.f. $\frac{1}{2}$; $x = 0$
   (d) Stretch parallel to y axis of s.f. $\frac{1}{3}$, or stretch parallel to x axis of s.f. $\sqrt{3}$; $x = 0$;
   (e) Translation $\begin{pmatrix} 3 \\ -5 \end{pmatrix}$; $x = 3$ (f) $y = (x - 1)^2 - 1$: translation $\begin{pmatrix} 1 \\ -1 \end{pmatrix}$; $x = 1$
   (g) $y = (x - 2)^2 - 1$: translation $\begin{pmatrix} 2 \\ -1 \end{pmatrix}$; $x = 2$
   (h) $y = 2\left[(x + 1)^2 - 1\frac{1}{2}\right]$: translation $\begin{pmatrix} -1 \\ -1\frac{1}{2} \end{pmatrix}$ then stretch parallel to y axis of s.f. 2; $x = -1$
   (i) $y = 3\left[(x - 1)^2 - 1\frac{5}{3}\right]$: translation $\begin{pmatrix} -1 \\ -\frac{5}{3} \end{pmatrix}$ then stretch parallel to y axis of s.f. 3; $x = 1$

2. (a) Translation $\begin{pmatrix} 90° \\ 0 \end{pmatrix}$ (b) Stretch parallel to x axis of s.f. $\frac{1}{3}$

(c) Stretch parallel to y axis of s.f. $\frac{1}{2}$ (d) Stretch parallel to x axis of s.f. 2

(e) Stretch parallel to x axis of s.f. $\frac{1}{3}$, and translation $\begin{pmatrix} 0 \\ 2 \end{pmatrix}$, in either order

3. (a) Translation $\begin{pmatrix} -60° \\ 0 \end{pmatrix}$ (b) Stretch parallel to y axis of s.f. $\frac{1}{3}$ (c) Translation $\begin{pmatrix} 0 \\ 1 \end{pmatrix}$

(d) Translation $\begin{pmatrix} 90° \\ 0 \end{pmatrix}$, then stretch parallel to x axis of s.f. $\frac{1}{2}$

4. (a) (i)

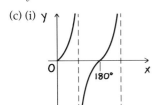

(ii) $y = \sin x$

(b) (i)

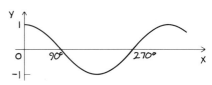

(ii) $y = \cos x$

(c) (i)

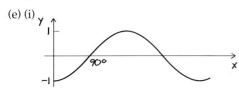

(ii) $y = \text{tax } x$

(d) (i)

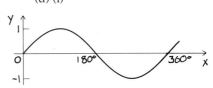

(ii) $y = \sin x$

(e) (i)

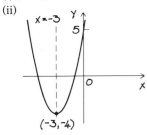

(ii) $y = -\cos x$

5. (i) $a = -4$

(ii)

6. $p = 3, q = 2$

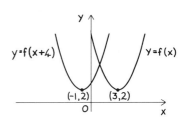

7. (a)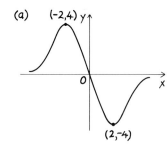

(b)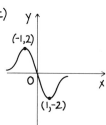

(c)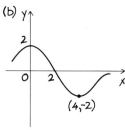

(d)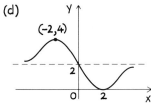

(e)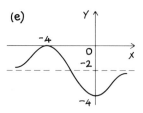

(f)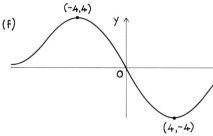

8. (i)

(ii) 21 machines   **9.** (ii) 100°C

Exercise 3C

1. (a) Stretch parallel to y axis of s.f. 2, then reflection in x axis, either order; $x = 0$

(b) Reflection in x axis then translation $\begin{pmatrix} 0 \\ 4 \end{pmatrix}$; $x = 0$

(c) $y = -(x-1)^2$: translation $\begin{pmatrix} 0 \\ 1 \end{pmatrix}$ and reflection in x axis, either order; $x = 1$

2. (a) (i)

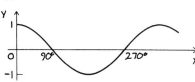

(ii) $y = \cos x$

(b) (i)

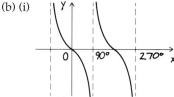

(ii) $y = -\tan x$

(c) (i)

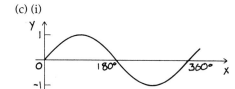

(ii) $y = \sin x$

(d) (i)

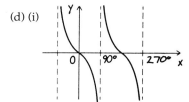

(ii) $y = -\tan x$

(e) (i)

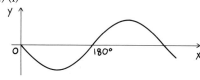

(ii) $y = -\sin x$

3. (i) $a = 3$, $b = 5$
(iii) $y = 6x - x^2 - 14$

(ii)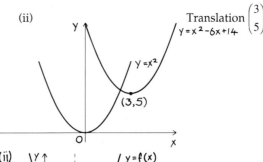
Translation $\begin{pmatrix} 3 \\ 5 \end{pmatrix}$
$y = x^2 - 6x + 14$

4. (i)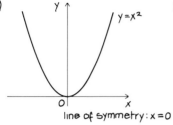
line of symmetry: $x = 0$

(ii)
$y = f(x)$
$(2,1)$
$x = 2$

(iii)
$(2,-1)$
$y = -f(x)$
$x = 2$

(iv)
$y = f(x+1) + 2$
$(1,3)$
$x = 1$

5. $a = 2$, $b = 1$, $c = 3$; $(-1,3)$

6. (a)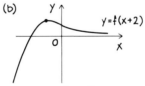
$(1,3)$
2
$y = f(x) + 2$

(b)
$y = f(x+2)$

(c)
$(\frac{1}{2},1)$
$y = f(2x)$

7.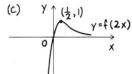

$\dfrac{x^2}{9} + \dfrac{y^2}{4} = 1$

8. (a) $y = f(x + 2)$ (b) $y = -f(x)$ (c) $y = f\left(\frac{x}{2}\right)$ (d) $y = f(x) - 3$ (e) $y = f(-x)$, (or $y = 2 - f(x)$) (f) $y = \frac{3}{2} f(x)$

Exercise 3D

1. (a) $8x^3$ (b) $2x^3$ (c) $(x + 2)^3$ (d) $x^3 + 2$ (e) $8(x + 2)^3$ (f) $2(x^3 + 2)$ (g) $4x$ (h) $\left[(x + 2)^3 + 2\right]^3$ (i) $x + 4$

2. (a) $f^{-1}(x) = \dfrac{x - 7}{2}$ (b) $f^{-1}(x) = 4 - x$ (c) $f^{-1}(x) = \dfrac{2x - 4}{x}$ (d) $f^{-1}(x) = \sqrt{(x + 3)}, x \geq -3$

3. (i),(ii)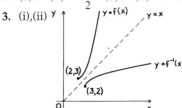
$y = f(x)$
$y = x$
$(2,3)$
$y = f^{-1}(x)$
$(3,2)$

4. (i) fg (ii) g^2 (iii) fg^2 (iv) gf

5. (i) $f(x)$ not defined for $x = 4$; $h(x)$ not defined for $x > 2$

(ii) $f^{-1}(x) = \dfrac{4x+3}{x}$; $h^{-1}(x) = 2 - x^2$, $x \geq 0$ (iii) $g(x)$ is not one-to-one.

(iv) Suitable domain: $x \geq 0$

(v) No: $fg(x) = \dfrac{3}{x^2 - 4}$, not defined for $x = \pm 2$; $\quad gf(x) = \left(\dfrac{3}{x-4}\right)^2$, not defined for $x = 4$.

6. (i) x (ii) $\dfrac{1}{x}$ (iii) $\dfrac{1}{x}$ (iv) $\dfrac{1}{x}$

7. $f^{-1} : x \to \sqrt[3]{\left(\dfrac{x-3}{4}\right)}$, $x \in \mathbb{R}$. The graphs are reflections of each other in the line $y = x$.

8. (i) $a = 3$

(iii) $f(x) \geq 3$

(iv) Function f is not one-to-one when domain is $\mathbb{R}$. Inverse exists for function with domain $x \geq -2$.

(ii)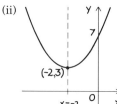

Exercise 3E

1. (a) Even (b) Odd (c) Neither (d) Neither (e) Odd (f) Even

2. (a) Even (b) Odd, periodic; $\frac{2}{3}\pi$ (c) None (d) Odd (e) Periodic; 2π (f) Odd, periodic; π

3. (i)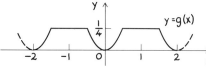

(ii) Half the period of $\sin x$

(iii) (a) 90° (b) 120° (c) 720°

4.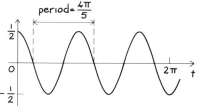

5. (i) $f(x) = x + 1$

(ii) $f(x) = 3 - x$

6.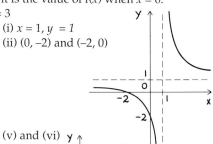

7.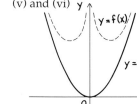

Exercise 3F

1. (a) A (−2, −6), since function is even. B (0,10) since it is the value of $f(x)$ when $x = 0$.

2. (i) $(0, -\frac{1}{4})$ (ii) $x = -2$, $x = 2$ and $y = 0$ **3.** $a = 1$, $b = 3$

(i) $x = 1$, $y = 1$

(ii) $(0, -2)$ and $(-2, 0)$

4. (i) $x = 0$

(v) and (vi)

(ii) Function is even
(iii) Always positive
(iv) Very close to $y = x^2$

5. (i) $a = -5, b = 2$

(ii)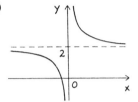

Exercise 4A

1. (a) $-\dfrac{4}{x^5}$ (b) $-20x^6$ (c) $-42x^{-7}$ (d) $-\dfrac{6}{x^3}$ (e) $-\dfrac{15}{x^6}$ (f) $-\dfrac{2}{x^2} + 3x^2$ (g) $-\dfrac{15}{x^4} + \dfrac{2}{x^2}$ (h) $\frac{1}{4}x^{-3/4}$ (i) $2x^{-3/5}$ (j) $\frac{1}{5}x^{-4/5}$

(k) $\frac{1}{2}x^{-1/2} - 3x^{-4} + 2$ (l) $\frac{4}{3}x^{1/3}$

2. (a) (i) $-2x^{-3}$ (ii) -128 (b) (i) $-x^2 - 4x^{-5}$ (ii) 3 (c) (i) $-12x^{-4} - 10x^{-6}$ (ii) -22 (d) (i) $12x^3 + 24x^{-4}$ (ii) 97.5

3. (i)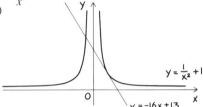

(ii) $\left(-\frac{1}{2}, 0\right)$

(iii) $-\dfrac{1}{x^2}$

(iv) -4

4. (i) $-\dfrac{8}{x^3} + 1$ (iii) 2 (v) 0 (vi) There is a minimum point at (2, 3).

5. (i)

$$y = \frac{1}{x^2} + 1$$

$$y = -16x + 13$$

(iii) $-\dfrac{2}{x^3}; -16$

(iv) The line $y = -16x + 13$ is a tangent to the curve

$$y = \frac{1}{x^2} + 1 \text{ at } (0.5, 5).$$

6. (i)

(ii) $\dfrac{1}{2\sqrt{x}}$

(iii) $\left(\frac{1}{16}, -\frac{3}{4}\right)$

(iv) No. The point $\left(\frac{1}{16}, -\frac{3}{4}\right)$ does not lie on the line $y = 2x - 1$.

7. (i) $3 + \dfrac{2}{x^3}$ (ii) 5 (iii) $y = 5x - 3$ **8.** (i) $2x - \dfrac{1}{x^2}$ (ii) 1 (iv) $(-2.4, 5.4), (0.4, 2.6)$

Exercise 4B

1. (a) $\dfrac{dy}{dx} = 3x^2; \dfrac{d^2y}{dx^2} = 6x$ (b) $\dfrac{dy}{dx} = 5x^4; \dfrac{d^2y}{dx^2} = 20x^3$ (c) $\dfrac{dy}{dx} = 8x; \dfrac{d^2y}{dx^2} = 8$

(d) $\dfrac{dy}{dx} = -2x^{-3}; \dfrac{d^2y}{dx^2} = 6x^{-4}$ (e) $\dfrac{dy}{dx} = \frac{3}{2}x^{1/2}; \dfrac{d^2y}{dx^2} = \frac{3}{4}x^{-1/2}$ (f) $\dfrac{dy}{dx} = 4x^3 + \dfrac{6}{x^4}; \dfrac{d^2y}{dx^2} = 12x^2 - \dfrac{24}{x^5}$

2. (a) $(-1, 3)$, minimum (b) $(3, 9)$, maximum (c) $(-1, 2)$, maximum and $(1, -2)$, minimum

(d) $(0, 0)$, maximum and $(1, -1)$, minimum (e) $(1, 2)$, minimum and $(-1, -2)$, maximum

(f) $(\sqrt{2}, 8\sqrt{2})$, minimum and $(-\sqrt{2}, -8\sqrt{2})$, maximum (g) $(16, 32)$, maximum

(h) $(-1, 2)$, minimum; $(-\frac{3}{4}, 2.02)$, maximum; $(1, -2)$, minimum.

3. (i) $4x(x + 2)(x - 2)$

(ii) $4(3x^2 - 4)$

(iv)

(iii) $(-2, -16)$, minimum; $(0, 0)$, maximum; $(2, -16)$, minimum

4. (i) Area $= xy = 18$ (ii) $P = 2x + y$ (iv) $\dfrac{dP}{dx} = 2 - \dfrac{18}{x^2}$; $\dfrac{d^2P}{dx^2} = \dfrac{36}{x^3}$ (v) $x = 3$ and $y = 6$

5. (i) $V = x^2y$ (ii) $A = x^2 + 4xy$ (iii) $A = x^2 + \dfrac{2}{x}$ (iv) $\dfrac{dA}{dx} = 2x - \dfrac{2}{x^2}$; $\dfrac{d^2A}{dx^2} = 2 + \dfrac{4}{x^3}$

(v) $x = 1$ and $y = \frac{1}{2}$

6. (i) $h = \dfrac{324}{x^2}$ (iii) $\dfrac{dA}{dx} = 12x - \dfrac{2592}{x^2}$; stationary point when $x = 6$ and $h = 9$

(iv) Minimum area $= 648\,\text{cm}^2$. Dimensions: $6\,\text{cm} \times 18\,\text{cm} \times 9\,\text{cm}$.

7. (i) $h = \dfrac{125}{r} - r$ (ii) $V = 125\pi r - \pi r^3$ (iii) $\dfrac{dV}{dr} = 125\pi - 3\pi r^2$; $\dfrac{d^2V}{dr^2} = -6\pi r$

(iv) $r = 6.45\,\text{cm}$; $h = 12.9\,\text{cm}$ (to 3 sig. figs.)

8. (i) $y = \dfrac{24}{x}$ (ii) $A = 3x + 30 + \dfrac{48}{x}$ (iii) $A = 54\,\text{m}^2$

9. (i) $h = \dfrac{1}{8r^2}$ (iii) $\dfrac{dA}{dr} = 2\pi r - \dfrac{\pi}{4r^2}$; $\dfrac{d^2A}{dr^2} = 2\pi + \dfrac{\pi}{2r^3}$; $r = \frac{1}{2}\,\text{m}$; $h = \frac{1}{2}\,\text{m}$

(iv) Minimum area $= \dfrac{3\pi}{4}\,\text{m}^2$ ($= 2.36\,\text{m}^2$ to 3 sig. figs.)

10. (i) Time taken $= \dfrac{100}{v}$ (ii) $C = 80v + \dfrac{200000}{v^2}$ (iii) $v = 17\,\text{kmh}^{-1}$ (iv) £2052 (to nearest £)

11. $r = 0.683\,\text{m}$ (to 3 sig. figs.) **12.** (i) $y = \dfrac{16}{x}$ (ii) $S = x^2 + \dfrac{256}{x^2}$ (iii) 32 (iv) $4\sqrt{2}\,\text{cm} = 5.66\,\text{cm}$ (to 3 s.f.)

Exercise 4C

1. (a) $3(x + 2)^2$ (b) $8(2x + 3)^3$ (c) $6x\,(x^2 - 5)^2$ (d) $15x^2\,(x^3 + 4)^4$ (e) $-3(3x + 2)^{-2}$

(f) $\dfrac{-6x}{(x^2 - 3)^4}$ (g) $3x(x^2 - 1)^{\frac{1}{2}}$ (h) $3\left(\dfrac{1}{x} + x\right)^2\left(1 - \dfrac{1}{x^2}\right)$ (i) $\dfrac{2}{\sqrt{x}}\left(\sqrt{x} - 1\right)^3$

2. (i) $9(3x - 5)^2$ (ii) $y = 9x - 17$

3. (i) $8(2x - 1)^3$ (ii) $(\frac{1}{2}, 0)$, minimum

(iii)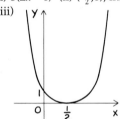

4. (i) $6x(x^2 - 4)^2$ (ii) $(0, -64)$, minimum; $(-2, 0)$ point of inflection; $(2, 0)$, point of inflection.

(iii)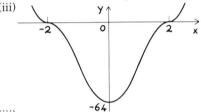

5. (i) $4(2x - 1)(x^2 - x - 2)^3$

(ii) $(-1, 0)$, minimum; $\left(\frac{1}{2}, \frac{6561}{256}\right)$, maximum; $(2, 0)$, minimum

(iii)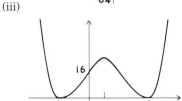

6. (i) $3x(3x - 2)\,(x^3 - x^2 + 2)^2$

(ii) $\dfrac{dy}{dx} = 0$ when $x = 1$ and when $x = 0$.

When $x < -1$ (e.g. 1.1) $\dfrac{dy}{dx} > 0$; when $-1 < x < 0$ (e.g. -0.5) $\dfrac{dy}{dx} > 0 \Rightarrow$ point of inflection at $x = -1$.

When x is just greater than 0 (e.g. 0.1) $\dfrac{dy}{dx} < 0 \Rightarrow$ maximum point at $x = 0$.

(iii) $a = \frac{2}{3}$ (iv) Gradient at $(1, 8)$ is 12; $y = 12x - 4$.

7. $\frac{1}{8}\,\text{km}$

Exercise 4D

1. (a) $x(5x^3 - 3x + 6)$ (b) $x^4(21x^2 + 24x - 35)$ (c) $2x(6x + 1)(2x + 1)^3$ (d) $-\dfrac{6}{(2x-1)^2}$

   (e) $-\dfrac{2}{(3x-1)^2}$ (f) $\dfrac{x^2(x^2+3)}{(x^2+1)^2}$

2. (i) $-\dfrac{1}{(x-1)^2}$ (ii) -1; $y = -x$ (iii) -1; $y = -x + 4$ (iv) The two tangents are parallel.

3. (i) $3x(x - 2)$ (iii)

   (ii) $(0, 4)$, maximum; $(2, 0)$, minimum.

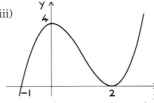

4. (i) $3(4x + 1)(x + 1)^2 (2x - 1)^2$ (ii) $x = -1$, point of inflection; $x = -\frac{1}{4}$, minimum; $x = \frac{1}{2}$, point of inflection.

   (iii) $P(-1, 0)$; $Q\left(-\frac{1}{4}, -\frac{729}{512}\right)$; $R\left(\frac{1}{2}, 0\right)$

5. (i) $\dfrac{\sqrt{x} - 2}{\left(\sqrt{x} - 1\right)^2}$ (ii) $\frac{1}{4}$ (iii) $(4, 8)$ (iv) Tangent: $y = 8$; normal: $x = 4$ (v) $Q\left(\frac{37}{4}, 8\right)$; $R(4, 29)$

6. (i) $-\dfrac{1}{(x-4)^2}$

   (ii) $4y + x = 12$

   (iii) $y = x - 3$

   (iv) $\dfrac{dy}{dx} \neq 0$ for any value of x

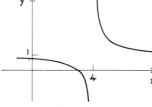

7. (i) $2\dfrac{(x+1)(x+2)}{(2x+3)^2}$ (ii) $(-1, -2)$; $(-2, -3)$ (iii) $(-1, -2)$, minimum; $(-2, -3)$, maximum.

Exercise 4E

1. (a) $-\frac{10}{3}x^{-3} + c$ (b) $x^2 + x^{-3} + c$ (c) $2x + \frac{1}{4}x^4 + \frac{5}{2}x^{-2} + c$ (d) $2x^3 + 7x^{-1} + c$ (e) $4x^{\frac{5}{4}} + c$ (f) $-\dfrac{1}{3x^3} + c$

   (g) $\dfrac{2x^{\frac{3}{2}}}{3} + c$ (h) $\dfrac{2x^5}{5} + \dfrac{4}{x} + c$ (i) $\dfrac{16x^{\frac{3}{4}}}{3} + c$

2. (a) $2\frac{1}{4}$ (b) $\frac{3}{4}$ (c) 41.4 (d) $-2\frac{2}{3}$ (e) 3.68 (f) $10\frac{2}{3}$

3. $21\frac{1}{3}$ 4. (i) $P(-4, 3)$; $Q(-2, 0)$; $R(2, 0)$; $S(4, 3)$ (ii) 8

5. (i) $(4, 2)$ (ii) and (iii)

   (iv) 6.77 (to 3 sig. figs.)

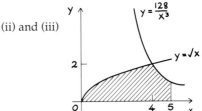

6. (ii) $B(1, -2)$ (iii) $1\frac{1}{4}$ 7. (i) $y = -\dfrac{2}{x} - 3x + c$ (ii) $y = -\dfrac{2}{x} - 3x + 17$ 8. $y = \frac{2}{3}x^{\frac{3}{2}} + c$; $y = \frac{2}{3}x^{\frac{3}{2}} + 2$

Exercise 4F

1. (a) $\frac{1}{4}(x + 1)^4 + c$ (b) $\frac{2}{3}(2x - 1)^{\frac{3}{2}} + c$ (c) $\frac{1}{8}(x^3 + 1)^8 + c$ (d) $\frac{1}{6}(x^2 + 1)^6 + c$ (e) $\frac{1}{5}(x^3 - 2)^5 + c$

   (f) $\frac{1}{6}(2x^2 - 5)^{\frac{3}{2}} + c$ (g) $\frac{1}{15}(2x + 1)^{\frac{3}{2}}(3x - 1) + c$ (h) $\frac{2}{3}(x + 9)^{\frac{1}{2}}(x - 18) + c$

2. (a) 205 (b) $928\,000$ (c) $5\frac{1}{3}$ (d) 30 (e) $222\,000$ (f) 586 (g) 18.1

3. (i) 4 (ii) -4; the graph has rotational symmetry about $(2, 0)$.

4. (i) 5.2 (ii) 1.6 (iii) 6.8 (iv) Because region B is below the x axis, so the integral for this part is negative.

5. (a) 4 (b) $2\frac{2}{3}$ (c) $22\frac{1}{2}$ (d) $1\frac{1}{9}$ 6. (i) $A(-1, 0)$: $x \geq -1$

7. (a) (i) $\dfrac{(1+x)^4}{4}+c$ (ii) $2\frac{2}{3}$

(b) $\frac{1}{3}\left(2\sqrt{2}-1\right)\approx0.609$

8.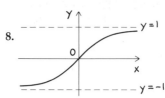

Area $=2(\sqrt{2}-1)\approx0.828$

Exercise 5A

1. $x=x_0e^{kt}$ 2. $t=\dfrac{1}{k}\ln\left(\dfrac{s_0}{s}\right)$ 3. $p=25e^{-0.02t}$ 4. $x=\ln\left(\dfrac{y-5}{y_0-5}\right)$

5. (i)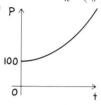

6. (i)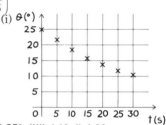

(ii) 100 (iii) 1218 (iv) 184 years

(ii) 25° (iii) 4.1° (iv) 22

7. (i)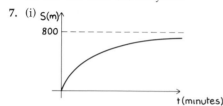

(ii) 621.5 m
(iii) 8.07 a.m. (to nearest minute.)
(iv) Never

8. (i)

(ii) $30\,\text{ms}^{-1}$, $8\,\text{ms}^{-1}$ (iii) $8.33\,\text{ms}^{-1}$ (iv) 8.7 s

Exercise 5B

1. (a) $\dfrac{3}{x}$ (b) $\dfrac{1}{x}$ (c) $\dfrac{2}{x}$ (d) $\dfrac{2x}{x^2+1}$ (e) $-\dfrac{1}{x}$ (f) $1+\ln x$ (g) $x(1+2\ln(4x))$

(h) $-\dfrac{1}{x(x+1)}$ (i) $\dfrac{x}{x^2-1}$ (j) $\dfrac{1-2\ln x}{x^3}$

2. (a) $3e^x$ (b) $2e^{2x}$ (c) $2xe^{x^2}$ (d) $2(x+1)e^{(x+1)^2}$ (e) $e^{4x}(1+4x)$

(f) $2x^2e^{-x}(3-x)$ (g) $\dfrac{1-x}{e^x}$ (h) $6e^{2x}(e^{2x}+1)^2$

3. (i) $0.108e^{0.9t}$ (ii) $0.108\,\text{mh}^{-1}$; $0.266\,\text{mh}^{-1}$; $0.653\,\text{mh}^{-1}$; $1.61\,\text{mh}^{-1}$

4. (i) $\dfrac{dy}{dx}=(1+x)e^x;\dfrac{d^2y}{dx^2}=(2+x)e^x$ (ii) $\left(-1,-\dfrac{1}{e}\right)$

5. (i) $\dfrac{dy}{dx}=2+\ln(x^2);\dfrac{d^2y}{dx^2}=\dfrac{2}{x}$ (ii) $\left(-\dfrac{1}{e},\dfrac{2}{e}\right)$, maximum; $\left(\dfrac{1}{e},-\dfrac{2}{e}\right)$, minimum.

6. (i) $\dfrac{e^x(x-1)}{x^2}$

(ii) $(1,e)$, minimum

(iii)

7. (i) $y=\ln x\Rightarrow\dfrac{dy}{dx}=\dfrac{1}{x}$; $y=x\ln x\Rightarrow\dfrac{dy}{dx}=1+\ln x$ 8. (i) $(1-x)e^{-x}$ (ii) $\left(1,\dfrac{1}{e}\right)$

Exercise 5C

1. (a) $3\ln|x|+c$ (b) $\frac{1}{4}\ln|x|+c$ (c) $\ln|x-5|+c$ (d) $\frac{1}{2}\ln|2x-9|+c$ (e) $\ln|x^2+1|+c$ (f) $\frac{1}{3}\ln|3x^2+9x-1|+c$

2. (a) $\frac{1}{3}e^{3x}+c$ (b) $-\frac{1}{4}e^{-4x}+c$ (c) $-3e^{-x/3}+$ (d) $4e^{x^3}+c$ (e) $-\dfrac{2}{e^{5x}}+c$ (f) $e^x-2e^{-2x}+c$

3. (a) $2(e^8-1)=5960$ (b) $\ln\frac{49}{9}=1.69$ (c) 0.018 (d) 4.70 (e) 0.906 (f) $\frac{1}{2}\ln\frac{29}{9}=0.585$

4. (i) $\frac{1}{2}(e-1)$ (ii) $\frac{1}{2}(e^4-1)$ (iii) $\frac{1}{2}(e+e^4)-1=27.7$ (to 3 sig. figs.)

5. (i) $(1-2x^2)e^{-x^2}$ (ii) $\dfrac{1}{\sqrt{2}}$ (iii) 0.294 6. $0.490;0.314$

7. (i) P(2,4); Q(–2,–4) (ii) 8.77; 14.2 (to 3 sig. figs.)

8. (i) $\frac{1}{2}(1 - e^{-x^2})$ (ii) 0.3161; 0.4908; 0.4999; 0.5000 (iii) $\frac{1}{2}$

9.

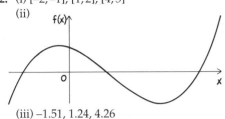

(ii) $\ln\left(\dfrac{e^2+1}{2}\right) \approx 1.434$ (iii) $\ln\left(\dfrac{e^2+1}{2}\right) \approx 1.434$

(iv) The same. The substitution transforms the integral in (ii) into that in (iii).

Exercise 6A

1. 1.62, 1.28

2. (i) [–2,–1]; [1,2]; [4,5]
(ii)

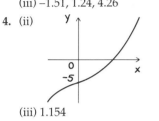

(iii) –1.51, 1.24, 4.26

3. (i) [1,2]; [4,5]
(ii) 1.857, 4.536

4. (ii)

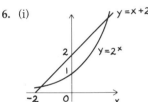

(iii) 1.154

5. (i) 2
(ii) [0,1]; [1,2]
(iii) 0.62, 1.51

6. (i)

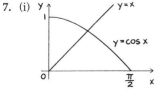

(ii) 2 roots (iii) 2, –1.690

7. –1.88, 0.35, 1.53

8. (a) (ii) No root (iii) Convergence to a non-existent root
(b) (ii) $x = 0$ (iii) Success
(c) (ii) $x = 0$ (iii) Failure to find root

Exercise 6B

1. (iii) 1.521 **2.** (iii) 2.120 **3.** (iii) $x = \sqrt[3]{(3-x)}$ (iv) 1.2134 **4.** (ii) 1.503

5. (i)

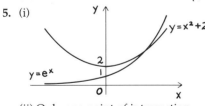

(ii) Only one point of intersection
(iv) 1.319 (to 3 dec. places)

6. (i)

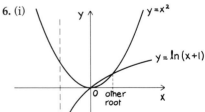

(ii) 0.747
(ii) 0.73909

7. (i)

Exercise 6C

1. (ii) −2.355 (iii) Eventually converges to −2.355
2. (i) $f(0) = 1$; $f(1) = -5$ (ii) 0.54 (iii) The process involves division by zero.
3. (ii) −0.46, 0.91, 3.73 (iii) $f(0) = 1$; $f'(0) = 1$ **4.** (ii) 1.86, 4.54 **5.** 0.567
6. (ii) 1.8019, 0.4450, −1.2470 (iii) No, e.g. −0.5 → 1.8019
7. (ii) −0.532, 0.653 (iii) 2.87938524157 (iv) After slow start, convergence is suddenly very rapid.
8. (i) 1 (ii)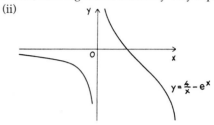

   (iii) 1.202